Fast Hopping Frequency Generation in Digital CMOS

Mohammad Farazian · Prasad S. Gudem
Lawrence E. Larson

Fast Hopping Frequency Generation in Digital CMOS

Mohammad Farazian
Qualcomm Incorporated
San Diego, CA
USA

Prasad S. Gudem
Qualcomm Incorporated
San Diego, CA
USA

Lawrence E. Larson
Brown School of Engineering
Brown University
Providence, RI
USA

ISBN 978-1-4899-9472-1 ISBN 978-1-4614-0490-3 (eBook)
DOI 10.1007/978-1-4614-0490-3
Springer New York Heidelberg Dordrecht London

Softcover reprint of the hardcover 1st edition 2013

Printed on acid-free paper

Springer is part of Springer Science+Business Media (www.springer.com)

To my parents
Mohammad Farazian

Preface

One of the challenges in implementing a frequency synthesizer for Multiband OFDM Ultra Wideband (MB-OFDM UWB) is overcoming the agility limitations of conventional synthesizers. The MB-OFDM proposal for UWB divides the spectrum from 3.1 to 10.6 GHz into 14 different bands, and frequency hops at the rate of 3.2 MHz between them with a specified frequency settling time of only 9.5 ns. Design techniques that eliminate the use of on-chip inductors, and which are compatible with low-voltage operation, are critical for increasing the level of integration for future implementations.

An inductor-less design methodology may have several advantages over traditional design techniques: (1) While the area required to implement an on-chip inductor does not scale down in the finer technology nodes, inductor-less designs benefit from technology scaling. (2) On the other hand, the quality factor of the on-chip inductors may worsen in finer technology nodes, which can lead to an increase in the required current consumption to generate a given voltage swing. (3) It is more straightforward to port an inductor-less design into a new technology node. The penalty for an inductor-less design methodology is a slightly increase in the current consumption to achieve the necessary gain and voltage swing in the absence of inductors.

In this work, a frequency plan is proposed that can generate all the required frequencies from a single fixed frequency and can implement any center frequency with a maximum of two levels of SSB mixing. In order to generate all the required frequencies for the operation of this frequency synthesizer out of a single frequency, fractional frequency dividers are needed. Therefore, a study is performed on the architectures that can obtain a fractional division ratio. This study involves an analysis of the operation, stability, and phase noise of injection-locked regenerative frequency dividers. In addition, the operation, stability, locking range, and phase noise of two-stage ring-oscillators, which are compact ways to generate quadrature output phases and can be used in injection-locked regenerative frequency dividers, are analyzed.

This work presents the first CMOS inductor-less single PLL 14-band frequency synthesizer for MB-OFDM UWB which is capable to perform any arbitrary band switching specified in less than 2 ns. Implemented in a 0.13 μm CMOS process, it uses a single 1.2 V supply voltage, and dissipates 135 mW. The mixing sideband level is better than −31 dBc and the phase noise is better than −110 dBc/Hz at 1 MHz offset.

University of California, San Diego, 2012

Mohammad Farazian
Lawrence E. Larson

Acknowledgments

I would like to thanks all my colleagues at University of California San Diego: Drs Himanshu Khatri, Alireza Kheirkhahi, Sunghwan Kim, Rahul Kodkani, Marcus Pan, Sanghoon Park, Junxiong Deng, Joe Jamp, Wingching Leung, Mehrdad Yazdani, and Mr. Yiping Han.

I would also like to thank Mr. Donald Kimball and Mr. Cuong Vu for their generous help by providing me with test and measurement facilities.

Lastly, I would like to acknowledge the support of the Center for Wireless Communications at UCSD and its member companies, and the University of California Discovery Grant.

Mohammad Farazian

Contents

Chapter 1
Introduction

1.1 Overview

The prevalence of universal serial bus (USB) technology in consumer households and beyond is unquestionable. USB has become a de facto standard for interfacing a personal computer (PC) to different peripherals such as webcams, printers, keyboards, and digital cameras. In addition, the fast speed of USB 2.0 (480 Mbps) makes it an appropriate medium for applications such as high-speed data transfer between a PC and external data storages or streaming high definition video contents. Indeed, at its arrival, the ubiquitous USB logo was a sign of comfort and relief. With the universal interface, consumers were liberated from the chains of interface limitations and USB related products soared. From computer pointing devices (mouse, keyboard, etc.) to digital imaging products, USB is everywhere.

However, the increase in USB related products encourages an environment filled with USB cables. Consumer households, for example, are typically cluttered with multiple digital cameras, portable MP3 players, and communication devices that interface with personal computers on a daily basis. As a result, dorm rooms, family rooms, and other personal areas are cluttered with USB cables, and users are trapped in an unnecessary web of tangles.

Indeed, a wireless USB solution will simplify a great deal of chaos, and the demand in developing a high-speed wireless personal area network (WPAN) to replace the existing USB cables with a wireless USB (WUSB) link is inevitable. Wireless USB, or WUSB, is an implementation of the USB standard using an appropriate WPAN standard that can match the data rate of USB in short range. WUSB devices must be capable of sending 480 Mbps at distances up to 2 m and 110 Mbps at up to 10 m. As WUSB is a relatively novel standard, it has yet to be fully adopted by the industry. However, support continues to grow as more devices are being ported to use the technology. Figure 1.1 shows application of a WPAN, in the form of WUSB, in home or office to connect PC to peripherals. Figure 1.2 shows the application of WPAN in consumer's houses to connect set top boxes, DVD players, Hi-Fi audio systems, digital cameras, and camcorders to high definition TV (HDTV) wirelessly.

M. Farazian et al., *Fast Hopping Frequency Generation in Digital CMOS*,
DOI: 10.1007/978-1-4614-0490-3_1,

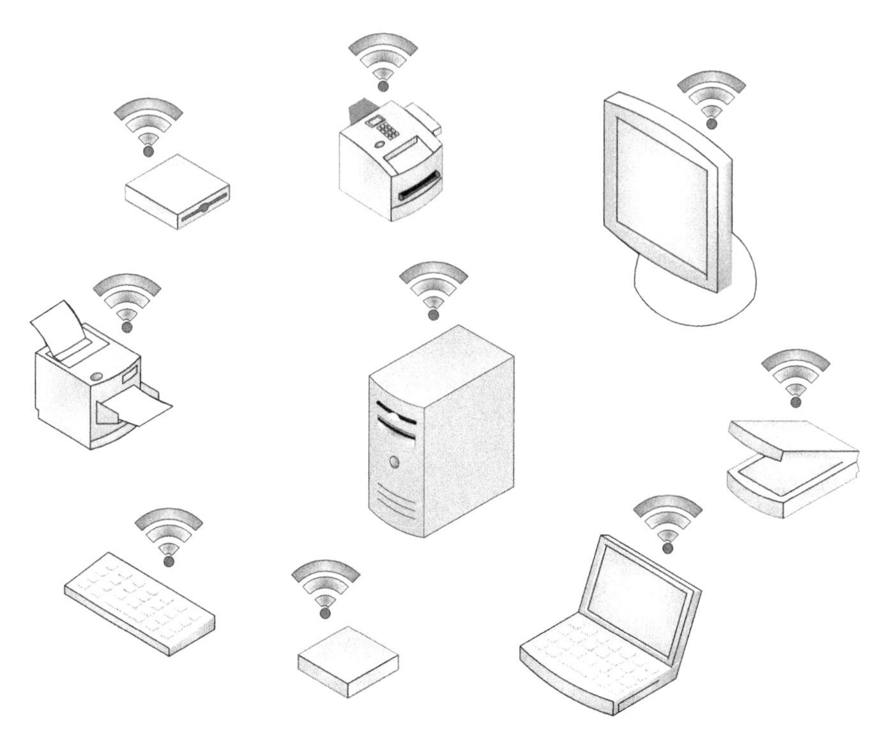

Fig. 1.1 Application of a WPAN to provide WUSB in home and office

1.2 Wireless Personal Area Networks

There are few existing standards for short range wireless data networks that could be used to provide a WPAN, including Bluetooth and ZigBee. Bluetooth operates at 2.4 GHz, uses GFSK modulation, and provides a data rate of 1 Mbps for short distances. An enhancement to Bluetooth (EDR) can increase this data rate to 3 Mbps.

ZigBee can operate at multiple carriers, including 2.4 GHz, and exploits direct-sequence spread spectrum (DSSS). The achievable data rate of ZigBee is limited to 250 kbps per channel and it can cover distances of 10–75 m.

In addition, there are other wireless data networks that are mainly designed to be used in a WLAN and they can provide moderately high data rates, including different variations of the Wi-Fi 802.11 standard. For instance, 802.11a and 802.11g operate at 5 and 2.4 GHz bands respectively, and occupy 20 MHz bandwidth, use 64 QAM constellation and OFDM modulation, and provide a maximum data rate of 54 Mbps, for a range of 100 m. Additionally, 802.11n can operate at both of the mentioned carrier frequencies and uses the same modulation, with bandwidth of 40 MHz, and provides up to 130 Mbps per beam, or a maximum of 600 Mbps when four beams are used, and covers a range of 300 m.

Fig. 1.2 Consumer electronics application of WPAN in homes

Lastly, ultra-wideband (UWB) can provide data rates as high as 480 Mbps for distances less than 2 m. This data rate drops to 110 Mbps for distances up to 10 m. More details about UWB will be discussed later in this chapter.

A comparison of the mentioned wireless data networks is depicted in Fig. 1.3.

Among these networks, only UWB and 802.11n can achieve the desired data rate required for applications such as wireless USB. However, it needs to be mentioned that UWB achieves this data rate with less power consumption and less hardware (single antenna vs. three or four beams), and with a lower cost. In addition, the coverage of UWB is more appropriate for a WPAN.

1.3 Multiband OFDM Ultra-Wideband

An ultra-wideband system is defined as a system that has a fractional bandwidth greater than 0.2, or a system that occupies an instantaneous bandwidth of greater than 500 MHz [1], and fractional bandwidth for any communication system is defined by

$$\mathit{Fractional\ Bandwidth} = (f_U - f_L)/f_C \qquad (1.1)$$

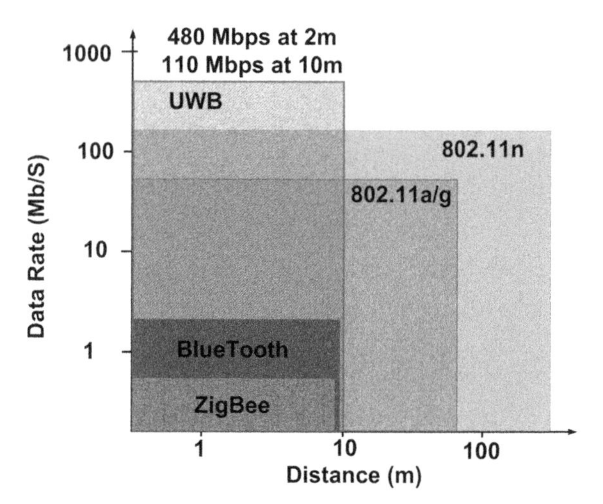

Fig. 1.3 Data rate as a function of distance for various wireless standards

where $f_C = (f_U + f_L)/2$ is the band center frequency, f_L is the band lower frequency, and f_U is the upper frequency. According to this definition, most communication standards are considered to be narrow band systems. For instance, the fractional bandwidth of Wi-Fi 802.11n that operates at 5.4 GHz and has a bandwidth of 40 MHz is only 0.0074.

The multiBand orthogonal frequency division modulation (MB-OFDM) scheme for Ultra-Wideband divides the spectrum of 3.1–10.6 GHz into 14 528-MHz bands. Each of the center frequencies for MB-OFDM can be expressed as

$$f_c = (5.5 + n) \times 528\,\text{MHz} \tag{1.2}$$

where n is the band number from 1 to 14. These fourteen bands are grouped into six Band Groups as shown in Table 1.1 [2]. All Band Groups consist of three bands except Band Group 5 which only includes bands 13 and 14.

The principles of operation of ultra-wideband can be described by the well-known Shannon–Hartley theorem for capacity of a communication channel with additive white Gaussian noise, as follows

$$C = B \log_2(1 + S/N) \tag{1.3}$$

where B is the bandwidth of the communication channel, and S and N are the average signal and noise power at the output of the channel, respectively [3].

According to (1.3) exploiting the large bandwidth enables UWB to achieve high data rates with low transmitted signal power and using simple constellations. This is in contrast to 802.11 a/g/n, which uses 64 QAM. In other words, UWB benefits from linear increase in the channel capacity by increasing the channel bandwidth, in contrast to the logarithmic increase of channel capacity due to the increase in the signal power.

Table 1.1 MB-OFDM Band Group allocation

Band group	Band ID n_b	Lower frequency (GHz)	Center frequency (GHz)	Upper frequency (GHz)
1	1	3.168	3.432	3.696
	2	3.696	3.960	4.224
	3	4.224	4.488	4.752
2	4	4.752	5.016	5.280
	5	5.280	5.544	5.808
	6	5.808	6.072	6.336
3	7	6.336	6.600	6.864
	8	6.864	7.128	7.392
	9	7.392	7.656	7.920
4	10	7.920	8.184	8.448
	11	8.448	8.712	8.976
	12	8.976	9.240	9.504
5	13	9.504	9.768	10.032
	14	10.032	10.296	10.560
6	9	7.392	7.656	7.920
	10	7.920	8.184	8.448
	11	8.448	8.712	8.976

Clearly UWB must work in the noise floor of other communication systems. This can be well seen from the indoor in-band emission limits of UWB which is 74 nW/MHz compared to Bluetooth and WLAN, which range from 1.2 to 30 μW/MHz (for different classes of Bluetooth) and 5–50 mW/MHz, respectively [4].

1.3.1 UWB Physical Layer (PHY) Description

Each of the bands in Table 1.1 consists of 128 subcarriers with a spacing of 4.125 MHz. A total of 122 subcarriers, including 100 data subcarriers, 10 guard subcarriers, and 12 pilot subcarriers to enable coherent detection, are used to transmit information in MB-OFDM UWB.

A combination of frequency-domain spreading, time-domain spreading, and forward error correction (FEC) coding are used to optimize the performance of MB-OFDM for a variety of channel conditions. A convolutional code with coding rates of 1/3, 1/2, 5/8, and 3/4 is used in forward error correction [2].

Next, a time-frequency code (TFC) is used to distribute the coded information over different bands within a group. There are three different types of time frequency codes defined in [2]. The first one is time-frequency interleaving (TFI) where the coded data is spread over three bands of a Band Group. The second one is two-band time-frequency interleaving (TFI2) where the coded data is spread over two bands

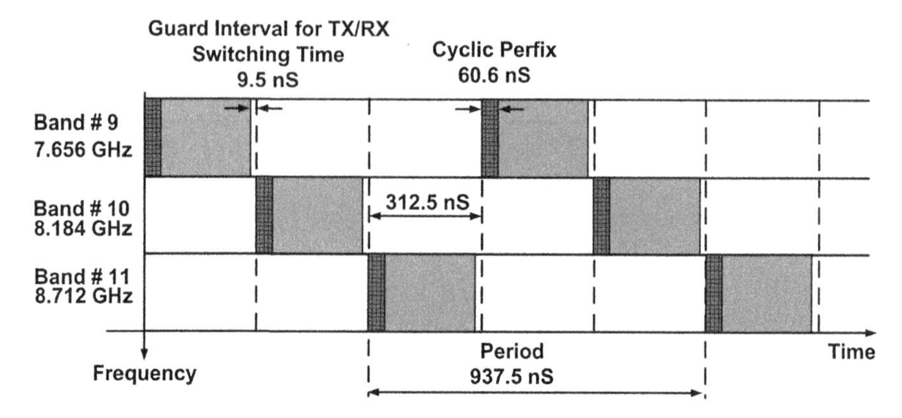

Fig. 1.4 An example of time-frequency interleaving for Band Group 6

Table 1.2 Time-frequency code patterns for Band Group 1

TFC number	Base sequence/ preamble	Band ID (n_b) for TFC					
1	1	1	2	3	1	2	3
2	2	1	3	2	1	3	2
3	3	1	1	2	2	3	3
4	4	1	1	3	3	2	2
5	5	1	1	1	1	1	1
6	6	2	2	2	2	2	2
7	7	3	3	3	3	3	3
8	8	1	2	1	2	1	2
9	9	1	3	1	3	1	3
10	10	2	3	2	3	2	3

within a Band Group, and the third is referred to as fixed frequency interleaving (FFI) where the coded data is transmitted over a single band of a Band Group.

An example of time-frequency interleaving, also known as band switching or frequency hopping, for bands in Band Group 6 is shown in Fig. 1.4.

Time frequency codes (TFC) required for different Band Groups are specified in [2]. TFC codes of Band Group One are shown in Table 1.2. This table shows a total of ten TFC codes, hence it can support up to ten different channels. Time frequency codes for other Band Groups are similar to Band Group One, with the exception of the fifth and the sixth Band Groups. The Band Group 5 only consists of two bands (bands 13 and 14). Hence, it only has one two-band time-frequency interleaving (TFI2) channels and one fixed frequency interleaving (FFI) channel. Band Group Six shares one band with Band Group 3 and two bands with Band Group 4. As a result, all of its FFI codes and one of its TFI2 codes overlap with channels of Band Groups 3 and 4.

Table 1.3 Timing parameters of MB-OFDM UWB

Parameter	Value
Sampling frequency (f_s)	528 MHz
Number of subcarriers or FFT size (N_{FFT})	128
Number of data subcarriers (N_D)	100
Number of pilot subcarriers (N_P)	12
Number of guard subcarriers (N_G)	10
Number of subcarriers used ($N_T = N_D + N_P + N_G$)	12
Subcarriers frequency spacing ($\Delta f = f_s/N_{FFT}$)	4.125 MHz
FFT and IFFT period ($T_{FFT} = 1/\Delta f$)	242.42 nS
Symbol interval (T_{SYM})	312.5 nS
Symbol rate ($F_{SYM} = 1/T_{SYM}$)	3.2 MHz
PHY band switching time	9.47 nS

1.3.2 Signal Generation

The transmitted RF signal for MB-OFDM UWB can be described as

$$S_{RF}(t) = \text{Re}\left\{ \sum_{n=0}^{N_{packet}-1} S_n(t - nT_{SYM}) e^{j2\pi f_c(q(n))t} \right\} \tag{1.4}$$

where $S_n(t)$ is the complex baseband OFDM signal for the nth symbol ($s_n(t) = 0$ for $t \notin [0,\ T_{SYM}]$), T_{SYM} is the symbol length, N_{packet} is the number of symbols in each packet, $f_c(\cdot)$ is one of the MB-OFDM carrier frequencies, and $q(\cdot)$ is the mapping function that maps the nth symbol to an appropriate carrier frequency.

$S_n(t)$ is an OFDM signal generated from the complex coefficients of the nth symbol ($d_n(k)$) as shown below.

$$S_n(t) = \sum_{k=0}^{N_{\text{FFT}}-1} d_n(k) e^{j2\pi k \Delta f t} \tag{1.5}$$

where the $d_n(k)$ coefficients could either be data symbols, pilots, or other training sequences. The timing parameters of MB-OFDM are shown in Table 1.3.

MB-OFDM uses quadrature phase-shift keying (QPSK) modulation for the data rates 200 Mbps and lower and it uses dual carrier modulation (DCM) for the data rates of 320, 400, and 480 Mbps. Table 1.4 shows the data rate dependent modulation parameters of MB-OFDM UWB.

Table 1.4 Data rate-dependent modulation parameters of MB-OFDM UWB

Data rate (Mbps)	Modulation	Coding rate (R)	FDS	TDS	Coded bits/ 6 OFDM Sym.	Info bits/ 6 OFDM Sym.
53.3	QPSK	1/3	Yes	Yes	300	100
80	QPSK	1/2	Yes	Yes	300	150
106.7	QPSK	1/3	No	Yes	600	200
160	QPSK	1/2	No	Yes	600	300
200	QPSK	5/8	No	Yes	600	375
320	DCM	1/2	No	No	1200	600
400	DCM	5/8	No	No	1200	750
480	DCM	3/4	No	No	1200	900

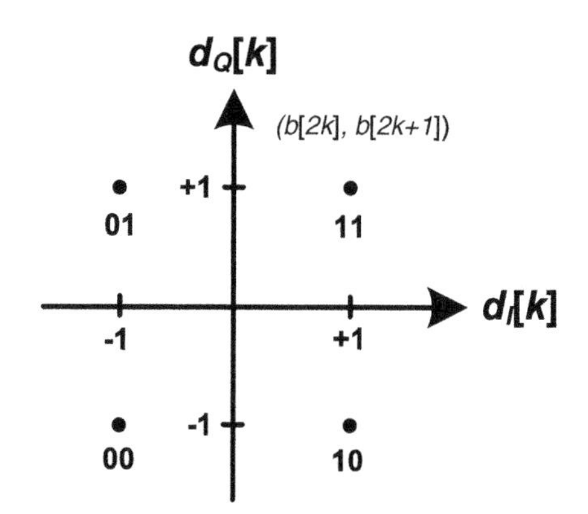

Fig. 1.5 Mapping for QPSK

It is customary to present the baseband binary interleaved coded serial data, defined in [2], by $b[k]$, where $k \in \mathbb{N}_0$, and $\mathbb{N}_0$ is the set of non-negative integer numbers. In this case, by using $\tilde{b}[k] = 2b[k] - 1$ we obtain a ± 1 bit stream. This ± 1 bit stream ($\tilde{b}[k]$) is used to generate QPSK, or DCM symbols. QPSK symbols are generated by transforming each two adjacent bits into a normalized complex number using a gray-coded mapping, as shown below.

$$d[k] = K_{MOD_Q}(\tilde{b}[2k] + j\tilde{b}[2k+1]) \tag{1.6}$$

where $K_{MOD_Q} = 1/\sqrt{2}$. The resultant QPSK constellation is shown in Fig. 1.5.

As discussed earlier, dual carrier modulation (DCM) must be used for data rates higher than 320 Mbps. To perform dual carrier modulation (DCM), the ± 1 bit stream of coded and interleaved binary serial baseband data ($\tilde{b}[k]$, $k \in \mathbb{N}_0$) are first grouped into groups of 200 bits. Later on, these 200 bits are grouped into 50 4-bit groups in the form of ($\tilde{b}[g(k)]$, $\tilde{b}[g(k)+1]$, $\tilde{b}[g(k)+50]$, $\tilde{b}[g(k)+51]$) where $k \in [0, 49]$ and

$g[k]$ is defined as follows

$$g[k] = \begin{cases} 2k & k \in [0, 24] \\ 2k + 50 & k \in [25, 49] \end{cases}. \tag{1.7}$$

Each 4-bit group is converted into two normalized complex numbers, $d[k]$ and $d[k+50]$, using the following transform [2, 4]:

$$\begin{aligned} d[k] = {} & K_{MOD_D}[(2\tilde{b}[g(k)] + \tilde{b}[g(k)+1]) \\ & + j(2\tilde{b}[g(k)+50] + \tilde{b}[g(k)+51])] \end{aligned} \tag{1.8a}$$

$$\begin{aligned} d[k+50] = {} & K_{MOD_D}[(\tilde{b}[g(k)] - 2\tilde{b}[g(k)+1]) \\ & + j(\tilde{b}[g(k)+50] - 2\tilde{b}[g(k)+51])] \end{aligned} \tag{1.8b}$$

where $K_{MOD_D} = 1/\sqrt{10}$. $d[k]$ and $d[k+50]$ would be placed on two different subcarriers. The DCM mapping of $d[k]$ and $d[k+50]$ are shown in Fig. 1.6a, b, respectively.

1.3.3 Coexistence, Emission Limits, and Worldwide Regulations of MB-OFDM UWB

Figure 1.7 shows the narrowband wireless systems that exist in the vicinity of the UWB span or are overlapping with that [4–6]. As can be seen from this figure, certain wireless systems, such as WiMax and different variations of Wi-Fi, overlap with the spectrum of 3.1–10.6 GHz. Operation of a UWB device must not cause any performance degradation for other devices that use the UWB spectrum. In addition, UWB must not generate out of band spurs that can affect operation of systems outside the UWB spectrum.

According to part 15 of federal communications commission (FCC) regulations, the allowable transmitted signal power of any UWB device needs to be less than −41.3 dBm in 1 MHz bandwidth while using a bandwidth greater than 500 MHz [1]. In addition, the maximum equivalent isotropically radiated power (EIRP) for the operation of indoor and hand held (outdoor) UWB devices are specified by FCC and is depicted in Fig. 1.8.

The worldwide regulation of the spectrum of 3.1–10.6 GHz is shown in Fig. 1.9 [2]. As can be seen from Fig. 1.9, this 7.5 GHz of bandwidth is unlicensed in the United States, but not available worldwide. Only Band Group 6 is available worldwide. As a result, all the UWB devices need to support Band Group 6. However, band 11 requires detection and avoid (DAA) for operation in Europe.

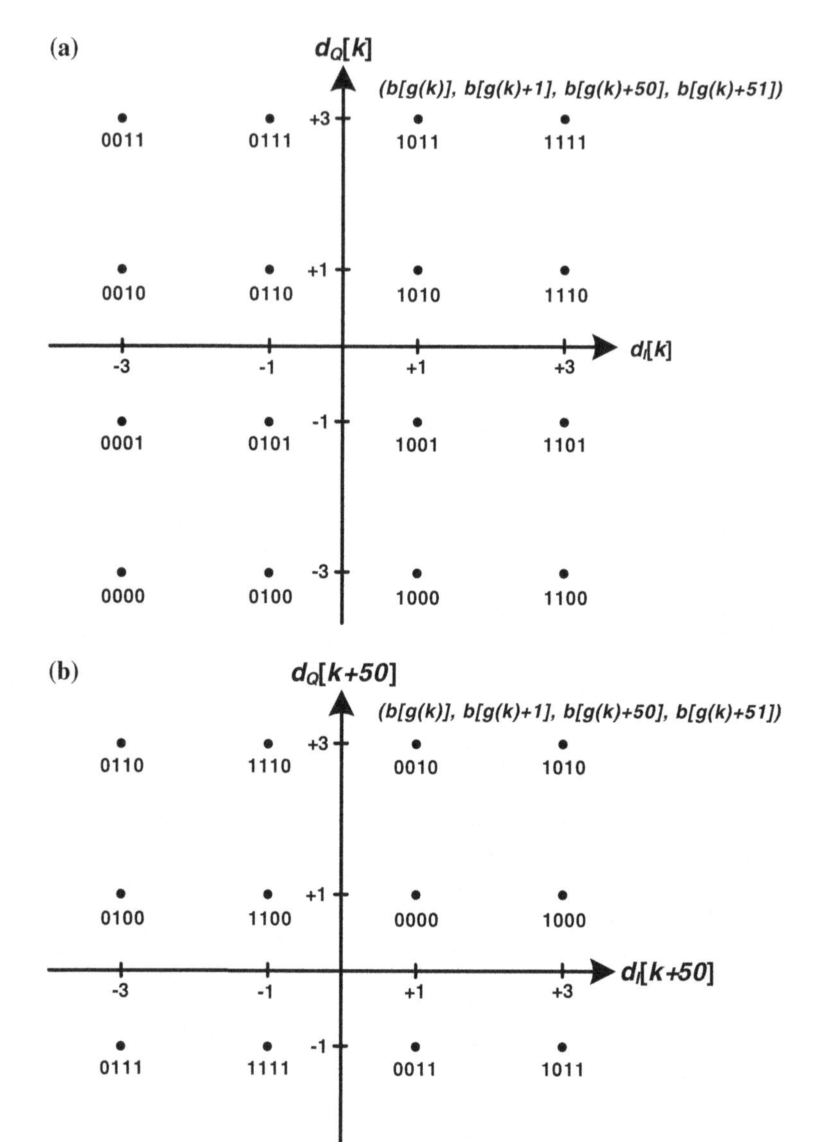

Fig. 1.6 DCM mapping **a** mapping for $d[k]$, **b** mapping for $d[k+50]$

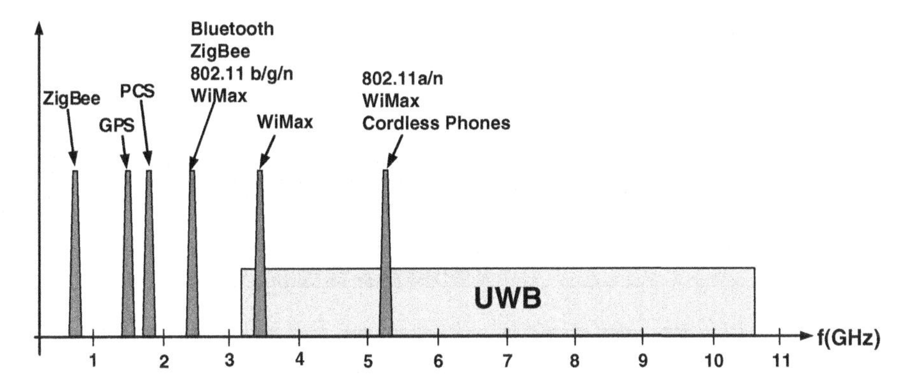

Fig. 1.7 Frequency usage of the spectrum of 1–11 GHz by wireless communication standards

Fig. 1.8 Emission limits for MB-OFDM UWB

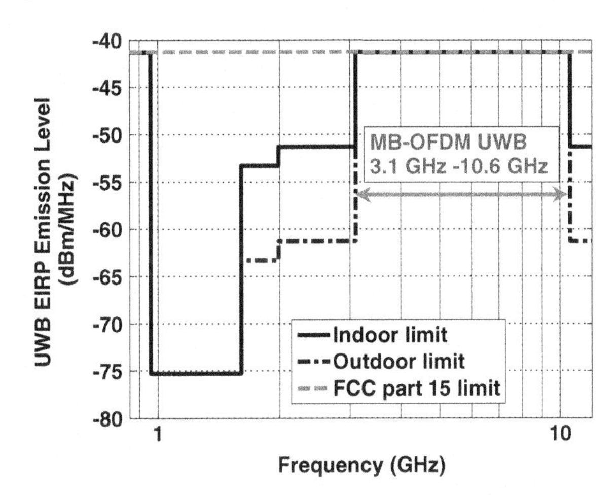

1.3.4 Detection and Avoid

In general, the operation of an ultra-wideband device must not causes any degradation in the quality of service of another wireless data networks, such as WiMax or Wi-Fi. As shown in Fig. 1.9, certain bands require detection and avoid (DAA) which means detecting activity of another communication service, a WiMax mobile terminal for instance, and avoiding any performance degradation caused by the operation of a UWB device. An example of destructive interference between UWB and WiMax in home applications is depicted in Fig. 1.10.

A detection technique based on fast fourier transform (FFT) to measure the LO leakage of a WiMax mobile terminal is presented in [7]. After detecting the activity of a WiMax mobile terminal, some actions need to be taken to avoid performance degradation of the WiMax terminal due to the activity of a UWB device. There are several options to avoid performance degradation of the detected mobile terminal

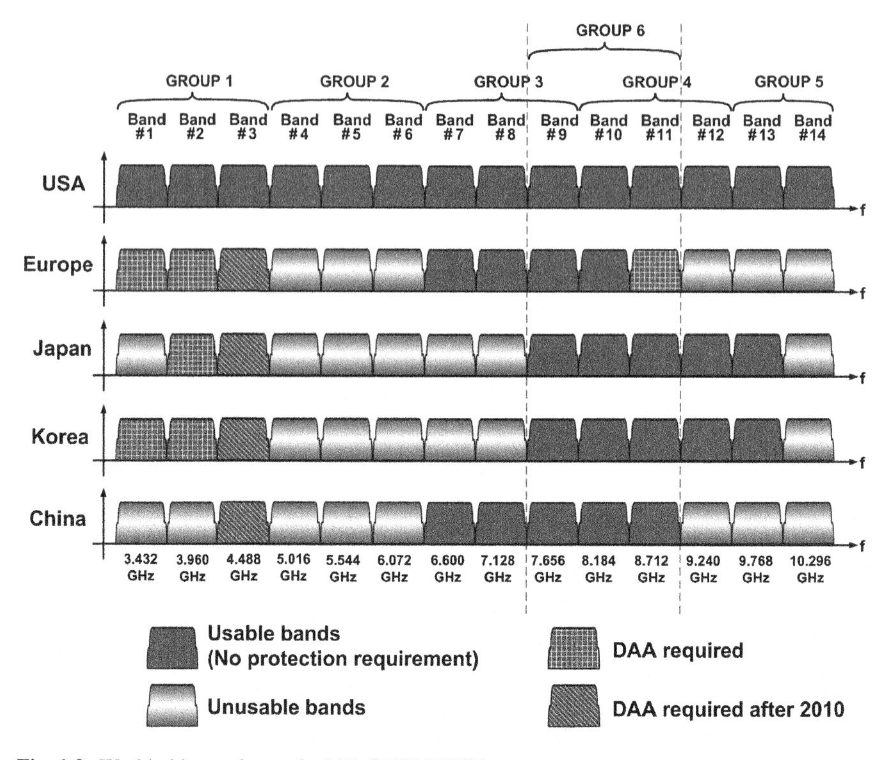

Fig. 1.9 Worldwide regulatory for MB-OFDM UWB

including controlling the transmit power of UWB device, frequency notching, and active interference control [8]. Any UWB device needs to provide support for transmit power control (TPC) to minimize the transmit power spectral density and yet providing a robust link to transfer data.

On the other hand, some mechanisms are provided in the PHY layer to allow nulling individual OFDM subcarriers [2]. These mechanisms along with the choice of the appropriate channels in each Band Group, allow some control of the spectrum and allow coexistence of UWB with other radios in that spectrum. An example of subcarrier nulling in MB-OFDM UWB is shown in [9].

1.4 Frequency Synthesizer for MB-OFDM UWB

A block diagram of a typical UWB transceiver is shown in Fig. 1.11. One of the main challenges in the design of a transceiver for UWB is to achieve a wide bandwidth with low power consumption and small die area.

One of the main challenges in the design of any UWB radio is to provide a hardware-efficient fast hopping frequency synthesizer that can implement the center

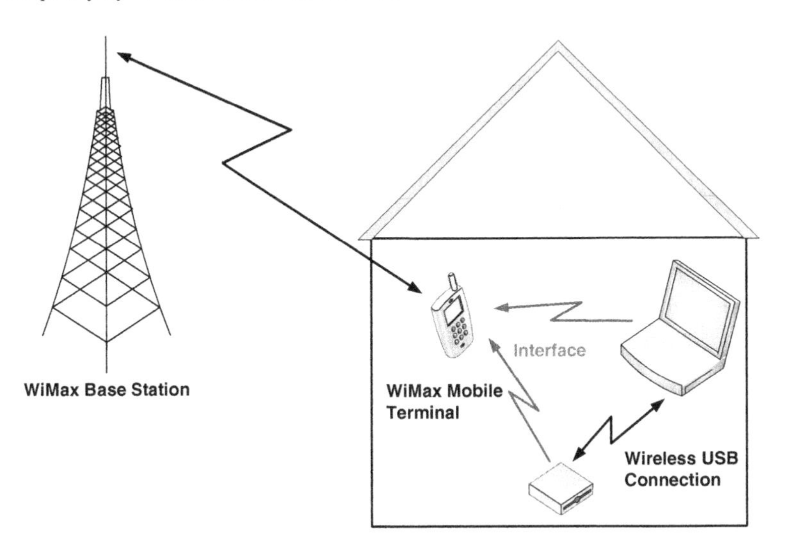

Fig. 1.10 An example of destructive interference in coexistence of WiMax and UWB

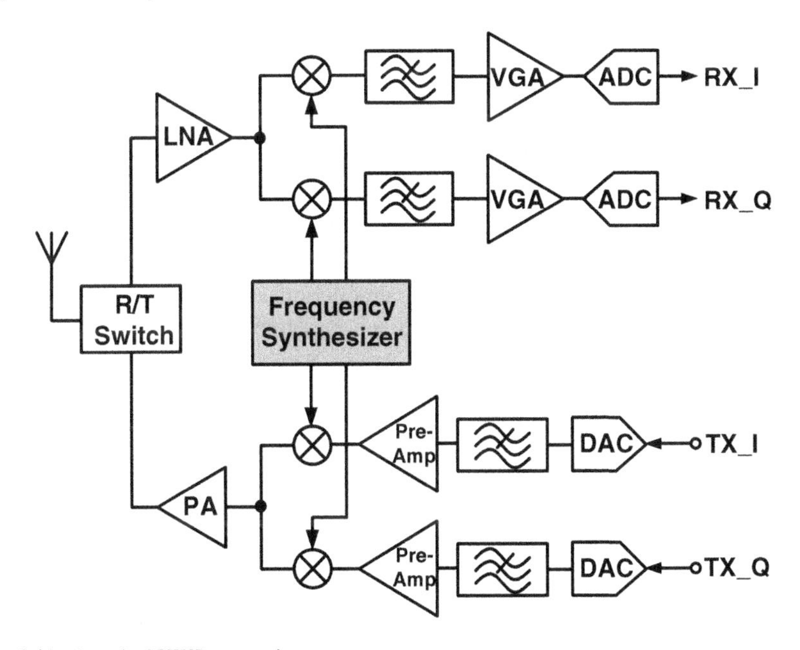

Fig. 1.11 A typical UWB transceiver

frequencies expressed by 1.2. As specified in Sect. 1.3, a UWB synthesizer frequency hops every 312.5 nS (hopping rate of 3.2 MHz) and the allotted time for a UWB synthesizer to perform the band switching is only 9.5 nS. This fast settling time poses several design challenges on the synthesizer for MB-OFDM UWB.

Moreover, to prevent bit error rate (BER) degradation, system simulations show that the in-band spurs must be as low as −30 dBc. Additionally, for coexistence with Wi-Fi, the spurious tones at 2.4 GHz and 5 GHz need to be smaller than −45 dBc and −50 dBc, respectively.

It is also revealed from system simulations that a phase noise of better than −100 dBc/Hz at 1 MHz offset is required. In addition, the transmitted center frequency needs to be accurate within ±20 ppm [2]. Furthermore, quadrature 50% duty cycle outputs are needed for proper operation of the SSB mixers of Fig. 1.11.

Figure 1.9 suggests that the MB-OFDM UWB frequency synthesizer needs to cover all 14 bands specified by 1.2 in order to have a universal UWB solution.

Meeting all these specification for the UWB frequency synthesizer in a digital CMOS technology using low power consumption and small chip area requires appropriate frequency planning and novel circuit design techniques. Accordingly, this challenge is the main focus of this book.

1.5 Book Organization

In this book, different architectures for implementing a frequency synthesizer for MB-OFDM UWB are studied, and an architecture suitable to implement a 14-band synthesizer for MB-OFDM UWB in digital CMOS technology is introduced. A hardware efficient implementation of a 14-band frequency synthesizer based on the architecture of Fig. 2.22 may require use of fractional frequency dividers as well as injection-locked frequency dividers (ILFDs). Therefore, this book contains some analysis of injection-locked frequency dividers and semidynamic regenerative frequency dividers which can be used to generate fractional frequency dividers. Moreover, quadrature signals are required for proper operation of the SSB mixers in this architectures.

An overview of the conventional frequency synthesis techniques and the hybrid architectures that can be used to implement a fast settle frequency synthesizer is presented in Chap. 2. This is followed by a proposed architecture for an inductor less fourteen- band CMOS frequency synthesizer for MB-OFDMUWB applications.

Chapter 3 includes an overview of quadrature generation at microwave frequencies using ring-oscillators. Furthermore, it consists of locking range comparison of multimodulus injection-locked frequency dividers that can combine the frequency division and quadrature generation. An injection-locked multimodulus four-stage ring-oscillator-based frequency divider is presented as an example. This frequency divider is implemented in 0.13 μm CMOS, and the measured data for locking range agrees with predictions.

A comparison of the techniques to implement frequency dividers with fractional division ratios is presented in Chap. 4. Moreover, this chapter presents a detailed study of the operation and stability of semidynamic regenerative frequency dividers for the first time, which provides a better understanding of the steady-state operation, locking range, and phase noise of this group of frequency dividers. For our analysis, driving the locking range and output phase noise of this frequency divider, required the locking range and free-running phase noise of the two-stage ring-oscillators. Therefore, some part of this chapter is devoted to study the free-running and injection-locked behavior of the two-stage ring-oscillator-based on negative-resistance delay cells. These analysis are presented for the first time. Moreover, a design example along with design considerations to obtain a 50 % quadrature output fractional division ratio is presented. All the calculations are compared with circuit simulations and show great agreement.

Different options to implement a cascade of two mixers, as well as the interface circuitry required to implement this cascade of mixers is studied in Chap. 5. This chapter also presents and overview of design techniques to implement a broadband amplifier in digital CMOS technology.

Chapter 6 presents the design of our proposed inductor-less, 14-band, fully quadrature frequency synthesizer for MB-OFDM UWB in a 0.13 μm CMOS technology. An inductor-less design methodology is introduced and techniques for spurious tones mitigation, such as SSB mixing, polyphase filtering, and low-voltage linearization techniques, in the absence of tuned inductive circuits are discussed. This chapter also contains of some analysis of the frequency response of multi-stage RC polyphase filters.

Finally, Chap. 7 concludes the this book, and presents some suggestion for further improvement of the performance metrics.

References

1. (2008) Federal Communications Commission, Revision of part 15 of the commissions rules regarding ultra-wideband transmission systems
2. Standard ECMA-368: High Rate Ultra Wideband PHY and MAC Standard, 3rd edition, http://www.ecma-international.org/ publications/standards/Ecma-368.htm, Dec. 2008
3. Shanmugam K (1985) Digital and Analog Communication Systems. Wiley, Chichester
4. Heidari G (2008) WiMedia UWB: technology of choice for wireless USB & Bluetooth. Wiley, Chichester
5. Siriwongpairat W, Liu KJR (2007) Ultra-wideband communications systems: multiband OFDM approach. Wiley-IEEE Press, New York
6. Jamp J (2007) Interference mitigation techniques for ultra-wideband systems. Ph.D. dissertation, University of California, San Diego
7. Park S, Larson LE, Milstein LB (2006) Hidden mobile terminal device discovery in a UWB environment, In: IEEE international conference on ultra-wideband, pp. 417–421
8. Aiello R, Batra A (2006) Ultra wideband systems: technologies and applications. Newnes, Burlington
9. Mishra S, ten Brink S, Mahadevappa R, Brodersen R (April 2007) Cognitive technology for ultra-wideband/WiMax coexistence. In: IEEE international symposium on new frontiers in dynamic spectrum access networks, pp 179–186

Chapter 2
Architectures for Frequency Synthesizers

2.1 Overview

This chapter starts with an overview of the conventional frequency synthesis techniques as well as the hybrid architectures that can be used to implement a fast settle frequency synthesizer. This is followed by a proposed architecture for an inductorless fourteen- band CMOS frequency synthesizer for MB-OFDM UWB applications.

2.2 Frequency Synthesis Techniques

The majority of frequency synthesis techniques fall into two categories: either direct frequency synthesis or indirect frequency synthesis. The direct frequency synthesis technique is based on using digital techniques to achieve fine frequency steps. While the achievable switching speed in this technique is fast, this technique is limited to lower frequencies due to the speed limitation of the digital circuits. Indirect frequency synthesis is based on using a phase-locked loop (PLL) to generate multiples (integer or non-integer) of a reference frequency. PLL-based techniques can synthesize higher frequency carriers; however, they obtain a slower settling and switching time. A combination of direct and indirect frequency synthesis techniques along with some other additional circuits, such as frequency mixers, can be used to achieve the required frequency of operation and switching speed in a frequency synthesizer. This section presents a detailed study of different architectures for a frequency synthesizer that can be used to implement a fast-settling fourteen-band frequency synthesizer that meets the specification stated in Sect. 1.4.

M. Farazian et al., *Fast Hopping Frequency Generation in Digital CMOS*,
DOI: 10.1007/978-1-4614-0490-3_2,

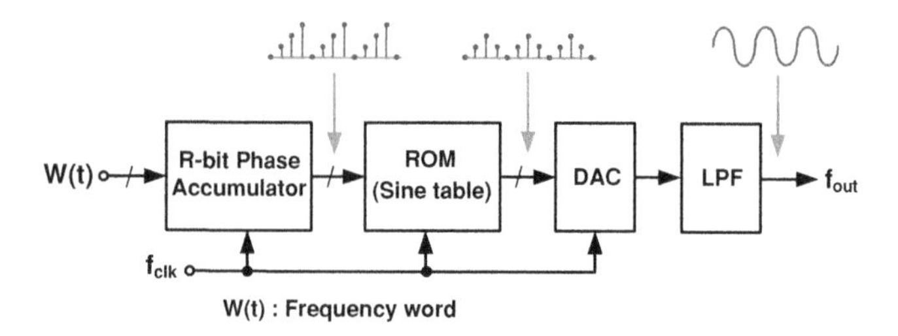

Fig. 2.1 Block diagram of a typical direct digital frequency synthesizer (DDFS)

2.2.1 Direct Digital Frequency Synthesizer

Direct digital frequency synthesis (DDFS) is a technique to synthesize frequencies and achieve a very fast settling time. A DDFS is composed of an accumulator, a ROM-based lookup table, and a digital-to-analog converter (DAC), and is usually, followed by a low-pass filter. A block diagram of a typical direct digital frequency synthesizer (DDFS) is shown in Fig. 2.1. The output of the accumulator is a discrete-time ramp representing phase information. As can be seen in Fig. 2.1, the accumulator is followed by a sine lookup table that converts the discrete-time ramp phase output of the accumulator into a discrete-time sine wave. Finally, the DAC and low-pass filter generate a continuous time waveform that is suitable for frequency translation applications. The oscillation frequency of the ramp at the output of the accumulator is [1]

$$f_{\text{out}}(t) = \frac{W(t) f_{\text{clk}}}{2^R} \tag{2.1}$$

where f_{clk} is the clock to the accumulator, $W(t)$ is the input to the accumulator, and R is the number of accumulator bits. As a result, based on the input frequency word ($W(t)$), the output of the accumulator is at a frequency that is a fraction of the master clock.

A block diagram of a (DDFS)-based frequency synthesizer for MB-OFDM UWB is shown in Fig. 2.2. It consists of a DDFS and a single-sideband (SSB) mixer for frequency translation. As discussed earlier, a DDFS can achieve a very fast frequency switching time. However, implementing a DDFS for UWB requires high-speed digital circuits and implementation of such high-speed logic is not very straightforward in current CMOS technology and also requires a very high power consumption. As a result, a hybrid architecture like the one shown in Fig. 2.2 is required to meet the frequency agility requirement of UWB and also to relax the design of the DDFS. Depending on the coverage of the UWB frequency synthesizer of Fig. 2.2, the DDFS needs to generate some integer multiples of the channel spacing (528 MHz), and the SSB mixer upconverts the output of the DDFS to the actual band center frequencies. Using an LO frequency that lies between two-band center frequencies slightly relaxes the DDFS coverage since it only needs to generate odd integers of 264 MHz (half the channel spacing) to achieve the same coverage. Although the architecture of Fig. 2.2

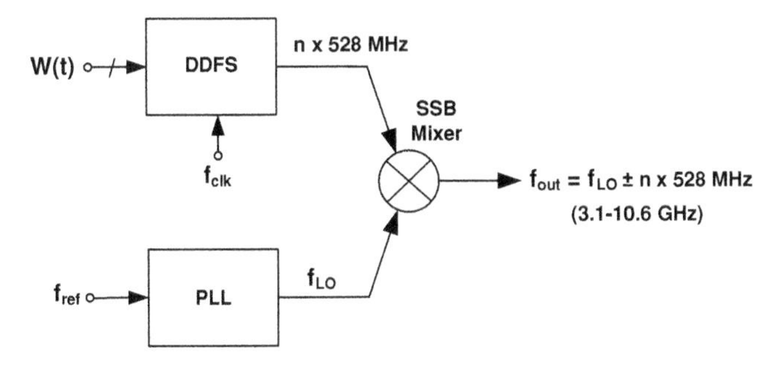

Fig. 2.2 A simplified block diagram of a DDFS-based frequency synthesizer for UWB

relaxes the speed requirements of DDFS, the accumulator, the ROM, and the DAC need to run at several giga hertz in order to generate three times the channel spacing at the output of the DDFS, which is challenging.

Another challenge is to mitigate the spurious tones at the output of the DDFS [1, 2]. Moreover, the architecture of Fig. 2.2 requires a linear SSB mixer to suppress the mixing spurious response at the output. Implementing a linear SSB mixer in a low-voltage CMOS process poses several challenging design constraints, and generally increases the power consumption. In addition, a SSB mixer requires quadrature phases of both mixing signals with sufficiently good amplitude and phase accuracy to achieve adequate side-band cancelation. So the DDFS needs to provide quadrature output phases. SSB mixers are integral parts of many different fast hopping frequency synthesizers, as will be discussed later in this chapter.

In summary, high-speed logic requirement makes the architecture of Fig. 2.2 less attractive, and also makes it very difficult for this architecture to cover a wide span of the UWB spectrum.

In special cases the DDFS of Fig. 2.1 can be simplified to a cascade of the ROM-based lookup table, DAC, and low-pass filter, which reduces the need to implement the accumulator. This can become more important at high frequencies where the implementation of high-speed logic is challenging. An example of this technique is presented in [3], where a simplified version of a DDFS is implemented by storing sine and cosine waveforms at frequencies of ± 264 and -792 MHz in two ROM-based lookup tables and selecting the appropriate frequency while using a fixed sampling rate of 4.224 GHz. The block diagram of this frequency synthesizer is shown in Fig. 2.3 [3]. As can be seen from Fig. 2.3, the outputs of the lookup tables go to the input of two four-bit current-steering DACs and low-pass filters. This frequency synthesizer uses a fractional-N PLL and a divide-by-two to generate the quadrature phases of 4.224 GHz. These signals are used in two SSB mixers to upconvert the outputs of the low-pass filters to the desired LO frequency for Band Group One. The frequency synthesizer of Fig. 2.3 is implemented in a 0.13 μm CMOS technology and performs frequency hopping within the center frequencies for Band Group one in 2 nS.

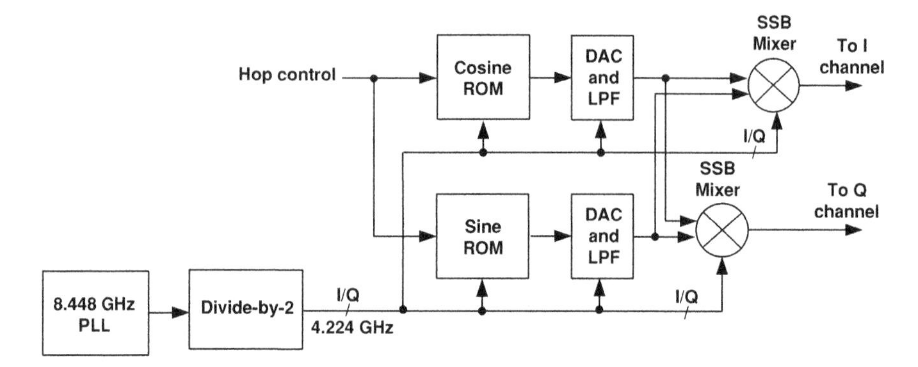

Fig. 2.3 Block diagram of the UWB frequency synthesizer of [3]

The ±264 MHz signal for the operation of the SSB mixer of the frequency synthesizer of Fig. 2.3 could be directly derived from the PLL output (8.448 GHz) using frequency dividers. However, generating the −792 MHz signal from the PLL output is not straightforward and requires another SSB mixer, or a frequency divider with a fractional division ratio. Using the DDFS to generate ±264 and −792 MHz has the advantage of suppressing the second and the third harmonics of these signals, while the remaining dominant fourth harmonic is suppressed by the low-pass filter at the output of the DAC in the DDFS [3]. In addition, the inductive tuned load in the SSB mixer and in the LO buffer after the SSB mixer provide further suppression of the remaining spurious tones.

2.2.2 Phase-Locked Loop-Based Approaches

A PLL is one of the most common ways to synthesize a frequency using a reference signal. Figure 2.4 shows a simplified block diagram of a PLL for frequency synthesis applications. The PLL of Fig. 2.4 consists of a phase-frequency detector (PFD), a charge-pump, a loop filter, and a voltage-controlled oscillator (VCO) in the forward path, and a frequency divider in the feedback path. The PLL aims to lock the phase of the feedback signal to the phase of the reference signal. Therefore, a phase detector is required to compare the phases of these two signals. However, a PFD can also be used to enable frequency detection as well [4]. A tri-state PFD is used in the block diagram of Fig. 2.4 [4, 5]. The output of the PFD determines the control voltage of the VCO through a charge-pump and loop filter. The PLL of Fig. 2.4 uses a second-order loop filter. Additional poles and zeros can be added to the loop filter to control the dynamic properties of the PLL as well as its stability. The frequency divider that is used in the feedback path indirectly leads to frequency multiplication at the output of the PLL, hence the output frequency of the PLL of Fig. 2.4 when phase locked is

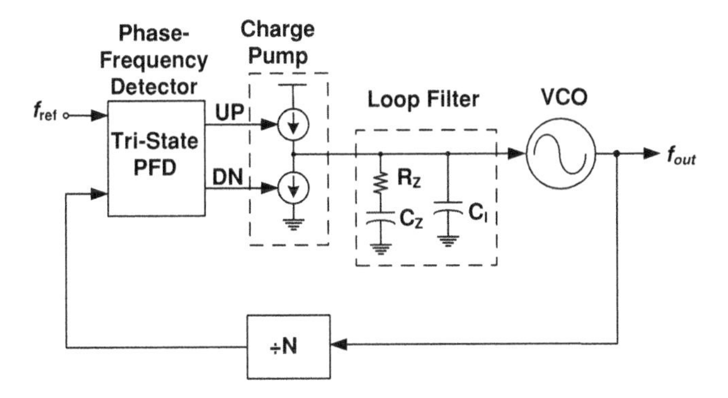

Fig. 2.4 Block diagram of a PLL-based frequency synthesizer with charge-pump

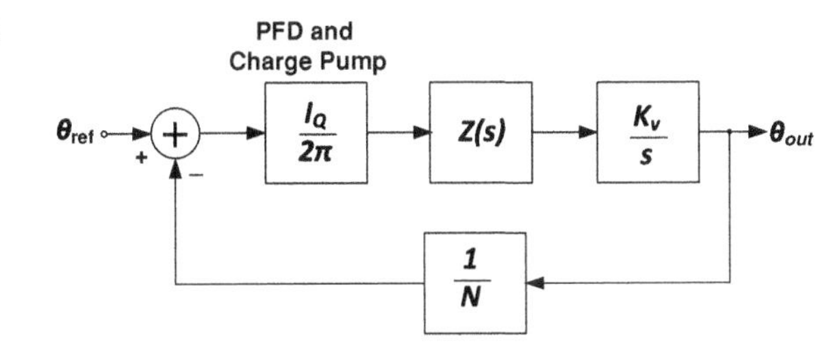

Fig. 2.5 Linear phase domain model for a charge-pump based PLL

$$f_{out} = N f_{ref} \tag{2.2}$$

where f_{ref} is the frequency of the reference frequency and N is the total division ratio used in the PLL.

To analyze the PLL of Fig. 2.4 we use a linear model for the phase of the input and output signals of the PLL. A linear model for the phase of the input and output signals of the PLL of Fig. 2.4 is shown in Fig. 2.5 [6]. The PLL of Fig. 2.4 tracks a phase step input if its loop gain has at least one integrator, and tracks a frequency step input if its loop gain has at least two integrators [7]. Therefore, a type-II PLL (a PLL with two poles at DC in its loop gain) is required to track a frequency with no steady-state error.

Based on this model, the loop gain of the PLL of Fig. 2.4 can be expressed by

$$G(s) = \frac{I_Q}{2\pi} Z(s) \frac{K_v}{s} \frac{1}{N} \tag{2.3}$$

where I_Q is the charge-pump current, $Z(s)$ is the impedance of the loop filter, K_v is the VCO gain, and N is the total division ratio in the loop.

For a type-II PLL the loop filter in its simplest form requires a pole at DC and a LHP zero for stability. However, for practical reasons it also needs a higher frequency pole to suppress the reference signal. This loop filter is shown in Fig. 2.4. The zero and the pole frequencies of the loop filter are chosen to achieve the desired loop

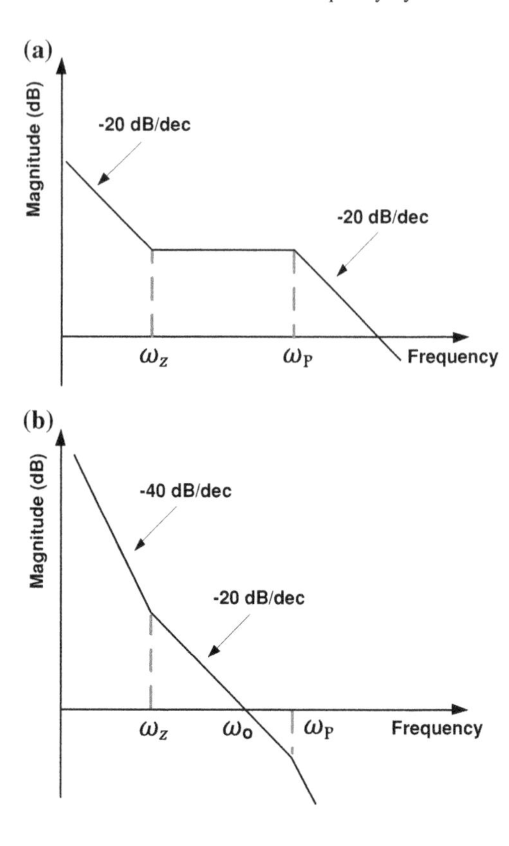

Fig. 2.6 **a** Magnitude of the transfer function of the loop filter used in the PLL of Fig. 2.4 and **b** magnitude of the loop gain of a charge-pump-based type-II PLL

bandwidth and phase margin. The impedance of the loop filter can be expressed by

$$Z(s) = \frac{1}{(C_Z + C_I)s} \cdot \frac{1 + R_Z C_Z s}{1 + R_Z (C_I || C_Z) s}. \tag{2.4}$$

In (2.4) C_I and C_Z are the shunt and the series capacitors in the loop filter and R_Z is the series resistor with C_Z to implement the zero of the impedance. As mentioned earlier, the impedance zero at the VCO input voltage improves the stability of the PLL, and is mainly determined by the required phase margin, while the location of the pole is determined by the required attenuation of the reference spurious tone. Therefore, the zero is chosen at a frequency that is a fraction of the unity gain frequency of the loop gain (ω_o) while the pole is chosen at a frequency higher than the unity gain frequency of the loop gain to minimize the effect of this pole on the phase margin of the PLL. The magnitude of the impedance of the loop filter and also the magnitude of the linear loop gain of the PLL of Fig. 2.4 are shown in Fig. 2.6a and b, respectively, where ω_Z is the zero of the impedance and is given by $1/R_Z C_Z$, ω_P is the pole of the impedance and is given by $1/R_Z(C_I || C_Z)$, and ω_o is the unity gain frequency of the loop.

It can be concluded from (2.4) that if $(\omega_Z/\omega_o)^2 \ll 1$ and $(\omega_o/\omega_P)^2 \ll 1$, the magnitude of the impedance of the loop filter in the vicinity of ω_o can be expressed by

$$|Z(j\omega_o)| \approx R_Z. \tag{2.5}$$

Consequently, from (2.3) and (2.5) the unity gain frequency of the transfer function of the loop gain of the PLL of Fig. 2.4 can be found as follows.

$$\omega_u \approx \frac{I R_Z K_v}{2\pi N} \tag{2.6}$$

The 3 dB bandwidth of the PLL of Fig. 2.4 depends on the unity gain frequency of its loop gain as well as its phase margin. For a phase margin of 60° the 3 dB bandwidth of the PLL of Fig. 2.4 is the same as the unity gain frequency of its loop gain and can be obtained from (2.6). In the design of Fig. 2.4, the loop filter parameters are chosen to obtain the desired phase margin and 3 dB bandwidth. The design and optimization of higher order loop filters are discussed in [8].

A wideband PLL is needed to meet the fast settling time requirement. But the bandwidth of a PLL also needs to be a fraction of the frequency of the reference signal due to stability reasons [6], and in practice this ratio is chosen to be no greater than 0.1. This can cause several practical issues. The first is settling time, and following approximation can be used to find the settling time of a PLL with a bandwidth of BW and a frequency step of f_{step} [9].

$$T_{\mathrm{settling}} = \frac{1}{BW\zeta} \ln\left(\frac{f_{\mathrm{step}}}{f_{\mathrm{error}}}\right) \tag{2.7}$$

In (2.7) ζ is the damping factor and is function of the phase margin, and f_{error} is the acceptable frequency error when PLL is settled [10]. For a phase margin of 50°, which chosen for fast settling [9], ζ is equal to five. Nonetheless, substituting a step frequency of 528 MHz and an frequency error of 1 kHz along with the optimal value of ζ in (2.7) leads to a reference frequency of greater than 2.8 GHz to obtain a settling time of 9.5 nS. However, in practice a slightly higher reference frequency is needed, since (2.7) is a linear approximation. To sum up, such a high reference frequency makes the implementation of the PFD or the phase detector very difficult. In addition, a reference frequency greater than the PLL step frequency mandates the use of a fractional-N PLL, which brings other challenges such as implementation of a delta-sigma modulator that operates at this reference frequency. These limitations make any single PLL frequency synthesizer approach unfeasible for these applications.

Several approaches have been suggested to overcome the stringent settling requirements of a PLL for UWB frequency synthesizer applications. One technique is based on using a dedicated PLL for each band, and using a multiplexer to select one of the PLL outputs, as shown in Fig. 2.7. In this technique each PLL settles to the desired carrier frequency of one of the bands at startup, and later the appropriate carrier frequency is selected by a multiplexer. Hence, there is no stringent settling requirement

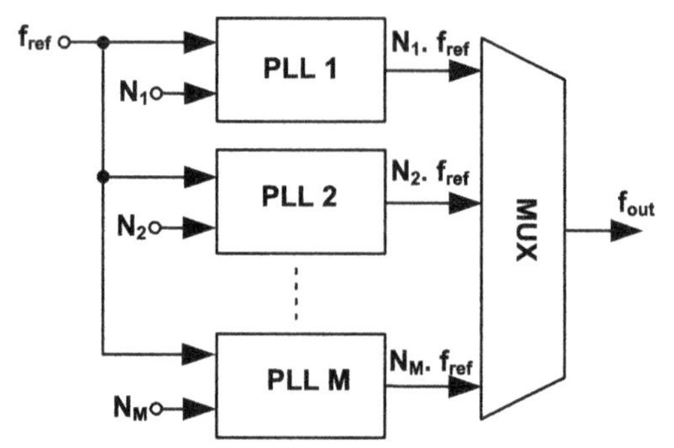

Fig. 2.7 Block diagram of a UWB frequency synthesizer based on multiple fixed-frequency PLLs

for the PLLs and the agility problem is reduced to the design of a multiplexer that performs the switching in less than 9.5 nS, which is a straightforward problem. An example of that is shown in Chap. 6. In addition, this technique does not require any SSB mixer, hence there is no mixer related linearity and spurious tone problem.

An example of this method to generate band center frequencies for Band Group One is presented in [11]. Other examples of this technique are the frequency synthesizers presented in [12, 13] that use three separate PLLs to generate band center frequencies for Band Groups one and three. Although this technique provides a solution for the agility problem of the conventional PLL, it usually leads to an increase in the die area and power consumption. Furthermore, using a dedicated conventional PLL for each band becomes very challenging for covering a larger number of bands due to practical issues of having several VCOs running simultaneously on one chip. The issues include VCO pulling and spurious tones due to different coupling mechanisms between the on-chip VCOs.

Another technique to overcome the agility requirement of UWB is based on using two fast-settling PLLs. In this technique while one PLL is providing the center frequency for that time slot, the other PLL is settling to the next carrier frequency. Consequently, the alloted time for PLL settling is increased to one symbol period, which is 312.5 nS. As a result, it is feasible to implement a phase detector and a charge-pump that operate at that slower speed, as shown in [9]. A simplified block diagram of this technique is shown in Fig. 2.8. As can be seen from this figure, this synthesizer does not require any SSB mixer, hence there is no mixer-related linearity and spurious tone issue.

An example of a UWB frequency synthesizer based on this technique is presented in [9] and is shown in Fig. 2.9. As can be seen from Fig. 2.9, this frequency synthesizer consists of two fast-settling PLLs, two frequency dividers (divide-by-two), and a four-to-one multiplexer. Although the frequency synthesizer of Fig. 2.9 generates seven band center frequencies for UWB, it is designed for an earlier frequency plan for MB-OFDM UWB [14] and only the PLL output frequencies after divide-by-two (3.432, 3.96, and 4.488 GHz) are compliant with the latest UWB frequency plan.

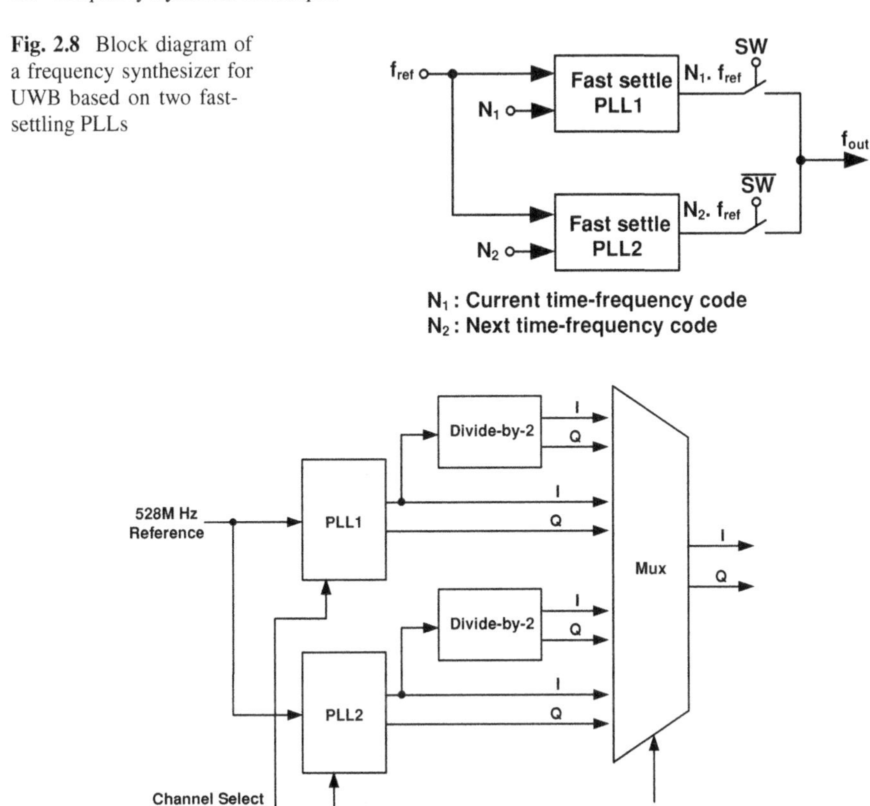

Fig. 2.8 Block diagram of a frequency synthesizer for UWB based on two fast-settling PLLs

Fig. 2.9 Block diagram of the UWB frequency synthesizer of [9]

The fast-settling PLLs used in the UWB frequency synthesizer of Fig. 2.8 are identical, and use a quadrature wideband VCO with a tuning range of 6.3–9.0 GHz, and a third-order on-chip active loop filter. Each PLL uses a reference frequency of 528 MHz, and is designed to achieve a bandwidth of 26 MHz with a phase margin of 50° for fast-settling [9]. The details of the circuit implementation of the PLLs of Fig. 2.8, including the design of the PFD and the charge-pump, are discussed in [9]. The frequency synthesizer of Fig. 2.9 was implemented in 0.18 μm CMOS technology and settles in 150 nS and achieves a phase noise of roughly −110 dBc/Hz at 1 MHz offset and sidebands of less than −52 dBc.

In short, the UWB frequency synthesizer of Fig. 2.8 overcomes the fast settling requirement for UWB and does not have the spurious tones of SSB mixer-based techniques. However, it is very difficult to expand this architecture to cover the entire UWB span since the VCOs cannot cover such a wide frequency range. On top of that, extending the frequency range of a VCO usually leads to degradation of its phase noise. Alternatively, multiple VCOs can be used to cover a wide range, but

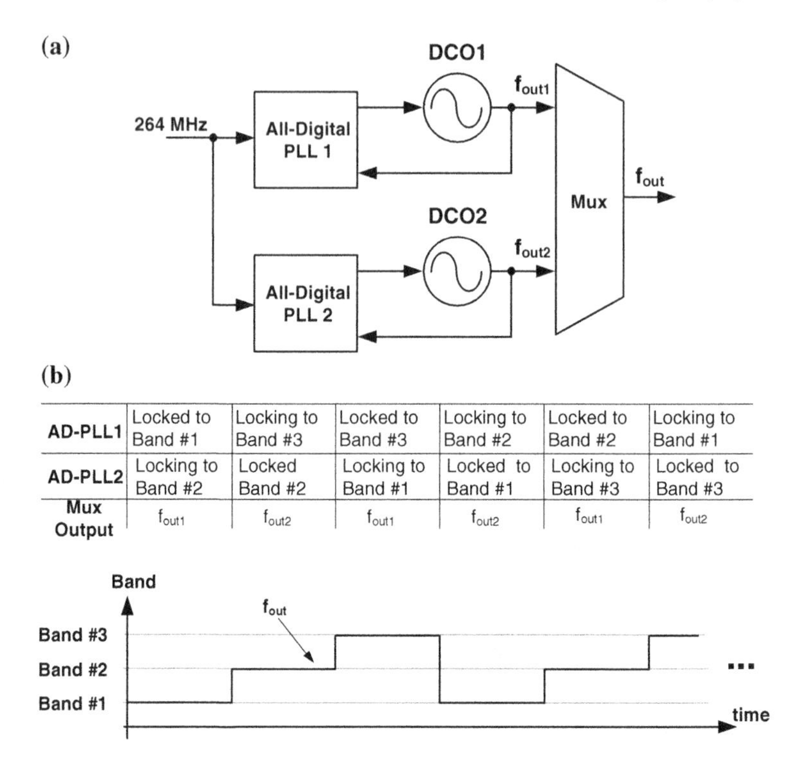

Fig. 2.10 **a** Block diagram of the frequency synthesizer used for each Band Group in the UWB frequency synthesizer of [15], **b** principles of operation of this frequency synthesizer in achieving fast frequency switching

this demands a larger die area. As a result, it is not very easy to extend the coverage of the frequency synthesizer of Fig. 2.8 to cover the entire UWB frequency range.

2.2.3 Digital Phase-Locked Loop-Based Approaches

Another PLL-based architecture to implement a fast agile frequency synthesizer for UWB is presented in [15]. This architecture is similar to what was presented in [9]. However, a more compact implementation of the frequency synthesizer is achieved using All-Digital PLLs (AD-PLL). In addition to covering Band Group Three in [9], Band Groups One and Six are also supported in the frequency synthesizer of [15]. This synthesizer is compliant with WiMedia v1.2 UWB PHY that requires operation in Band Groups One, Three, and Six.

Figure 2.10a shows the implementation of the synthesizer for each Band Group. As can be seen in Fig. 2.10a, two integer-N AD-PLLs are used to generate the center frequency for each band. While one PLL is providing the center frequency for that time slot, the other PLL is settling to the next carrier frequency. This technique increases the settling time budget for each PLL from 9.5 to 312 nS (time required for band switching) similar to what was discussed in the previous section. The outputs of two Digital PLLs are multiplexed and the appropriate output is selected at the time of frequency hopping. This switching can easily take place in less than 9.5 nS. The principles of operation of this frequency synthesizer within each Band Group is shown in Fig. 2.10b. The complete implementation of this UWB frequency synthesizer is shown in Fig. 2.11. As can be seen in this figure, the two-PLL frequency synthesizer of Fig. 2.10a is used for each Band Group and the entire frequency synthesizer consists of six AD-PLLs to cover three Band Groups. Using Digitally Controlled Oscillators (DCOs) helps to eliminate the inductors that are required inside VCOs. Each DCO consists of a cascade of four differential pair amplifiers and generates output signals with 45° phase shift, and achieves a phase noise of −84 dBc/Hz at 1 MHz offset, while dissipating 8 mA at the highest frequency of operation [15].

In addition to the digital PLLs, an analog PLL is used to generate the reference frequency for the digital PLLs as well as the clock for the operation of the digital baseband. The reference frequency of 264 MHz enables the use of integer-N PLLs to implement the digital PLLs of Fig. 2.11, since all the LO frequencies in the UWB spectrum are integer multiples of 264 MHz. In addition, the fact that the LO frequencies are odd integer multiples of the reference frequency simplifies the implementation of the Time-to-Digital Converter (TDC) required for the digital PLLs [15]. Finally, the relaxed accuracy and phase noise requirement of the synthesized LO frequencies (±20 ppm and −100 dBc/Hz at 1 MHz offset according to [16]) eases the implementation of digital PLLs. These features together facilitate a compact implementation of digital PLLs [15].

In spite of the poor phase noise performance of DCOs, using a wide PLL bandwidth helps to achieve acceptable phase noise for the operation of UWB radio.

Using multiple digital PLLs helps to eliminate the use of frequency dividers and SSB mixers to generate the center frequencies for different bands, as will be discussed in Sect. 2.2.8.

2.2.4 Sub-Harmonic Injection-Locking Techniques

One alternative way to overcome the PLLs settling time limitations in a UWB frequency synthesizer is by synchronizing an oscillator to an external signal [17, 18]. This synchronization can be done using a synchronization signal that has the same fundamental frequency as the target frequency, or by using a synchronization signal that is a subharmonic of the target frequency. The former approach is usually exploited to generate multiple phases of a carrier frequency to be used in mixers for frequency translation (up-conversion or down conversion), while the latter approach is often used for frequency multiplication, similar to using a PLL. It can be shown that in spite of the poor close-in phase noise of a free-running oscillator, an oscillator under injection-locking tracks the phase noise of the input source [19].

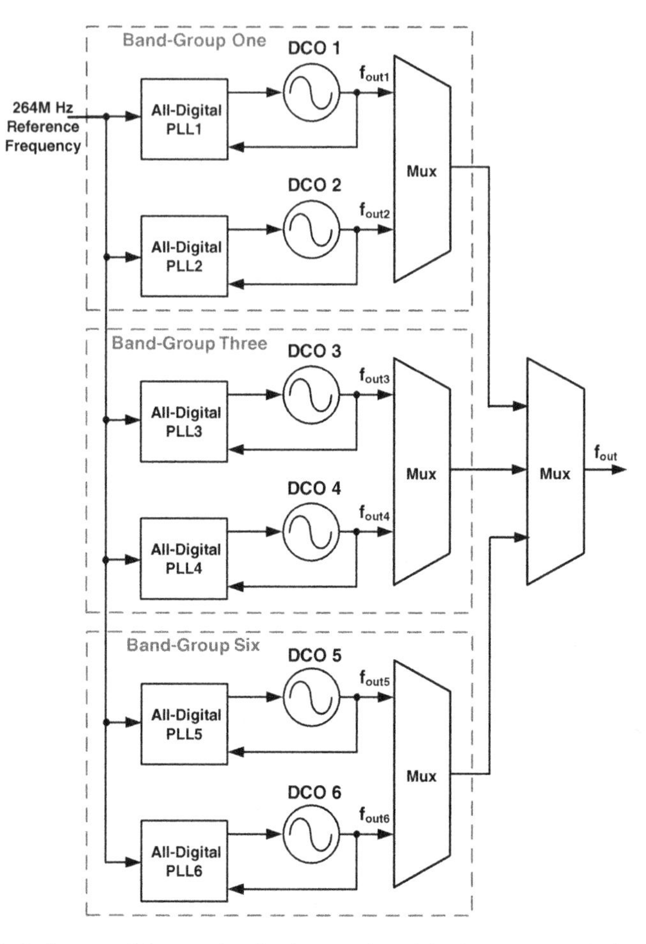

Fig. 2.11 Block diagram of the complete implementation of the frequency synthesizer of [15]

Therefore, to obtain acceptable phase noise, it is sufficient to use a clean synchronization signal. When this technique is used in a wideband frequency synthesizer the oscillator must have a wide enough tuning range. Alternatively, multiple oscillators need to be used to cover the entire frequency span. In this case, the settling time constitutes of two components: (1) the time required for tuning the free running frequency of the oscillator through coarse tuning or any other mechanism, and (2) the locking time required for synchronization of the oscillator to the injection signal [20]. As discussed in [20], a proper design can easily meet the settling time requirement for a UWB frequency synthesizer.

An example of a frequency synthesizer for UWB based on subharmonic synchronization, or injection-locking, is presented in [17, 20]. This frequency synthesizer uses an LC oscillator with a wide tuning range to generate the center frequencies for Band Group Six. In this frequency synthesizer, the oscillator can be injection-locked to a low phase noise reference signal that is a subharmonic of the desired frequency.

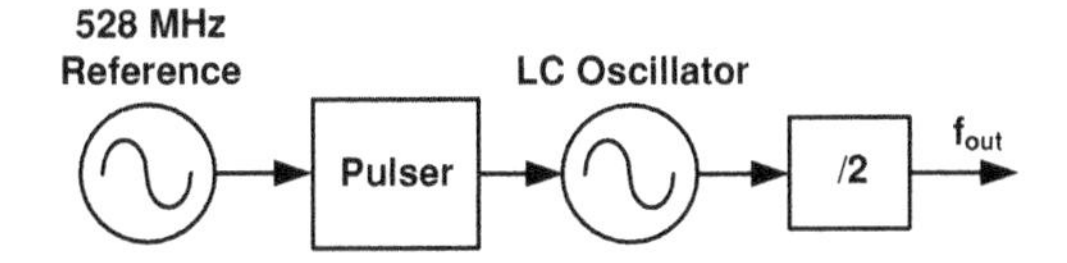

Fig. 2.12 Block diagram of a frequency synthesizers for UWB based on sub-harmonic injection-locking of an oscillator

A simplified block diagram of the frequency synthesizer of [17] is shown in Fig. 2.12. As can be seen from Fig. 2.12, an oscillator is followed by a divide-by-two to generate the quadrature output phases of the LO signal. Hence, the oscillator needs to oscillate at twice the frequency of the band center frequencies. In order to generate all the band center frequencies for Band Group Six, the fundamental frequency of the injection signal needs to be a common divisor of all these possible output frequencies. Considering the frequency synthesizer of Fig. 2.12, the greatest common divisor (GCD) for all the output frequencies is 528 MHz. Using a synchronization signal with a fundamental frequency that is the common divisor of all the output frequencies, but is not the GCD, can cause several practical issues: (1) a higher order harmonic of the fundamental frequency of this synchronization signal is needed to injection-lock the oscillator to the target desired frequency. It is more difficult to get the adequate power required for injection-locking at higher order harmonics. (2) Other harmonics will appear at the output through different coupling mechanisms, and will generate in-band spurious tones.

The synthesizer of [17] is implemented using an LC oscillator that uses capacitive coarse tuning to cover the frequency span required for Band Group Six. An external 528 MHz signal is used to generate the synchronization signal that injection-locks the oscillator. To create a synchronization signal that is rich in harmonics, this 528 MHz signal goes through a pulser circuit shown in Fig. 2.13. The synchronization can be done using both single-node and differential injection.

The frequency synthesizer of [17] achieves a very compact implementation on Silicon using a 90 nm CMOS technology and settles in less than 4 nS while it generates the band center frequencies for Band Group Six.

The frequency synthesizer of [18] uses similar principle for injection-locking a free-running oscillator. However, this frequency synthesizer uses a signal at the LO frequency or half the LO frequency for this purpose. Consequently, there is not an issue with spurious tones at the center frequencies of other UWB bands and it can achieve a much better spurious level. Figure 2.14 shows the top level block diagram of the frequency synthesizer of [18]. This frequency synthesizer generates the center frequencies for Band Groups One, Three, and Four. As can be seen from Fig. 2.14, a four-stage ring-oscillator is used to cover the entire frequency span of these three-Band Groups. Each delay cell consists of a differential pair and a programmable PMOS load through which the oscillator can be coarse tuned to different frequencies.

To achieve an acceptable phase noise at the output of the injection-locked ring-oscillator, the synchronization signal can be generated using a PLL with an LC-VCO. However, to overcome the fast settling time requirements, three separate PLLs are needed. These three PLLs are identical, and cover a frequency range of

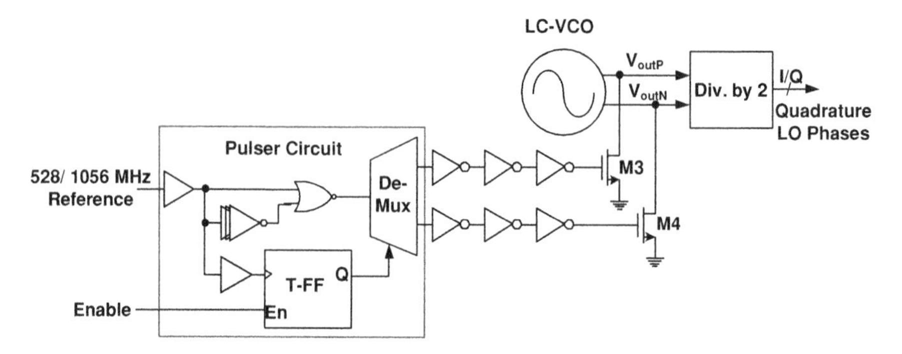

Fig. 2.13 Block diagram of the pulser used in the UWB frequency synthesizer of [17] to generate high harmonic contents of the injection signal

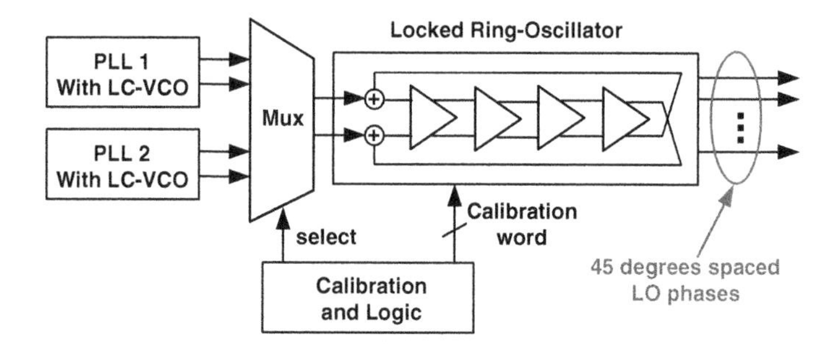

Fig. 2.14 Block diagram of the UWB frequency synthesizer of [18]

3.300–4.620 GHz with a frequency step of 132 MHz, and they use a reference frequency of 66 MHz. Each PLL generates one of the band center frequencies of Band Group One for operation in this Band Group. But for operation in Band Group Three or Band Group Four, the PLLs generate the synchronization signal at half the center frequency for each band. The outputs of these three PLLs are multiplexed and are injected to the output of the first delay cell of the ring-oscillator.

The frequency synthesizer of [18] requires generation of 45° spaced LO phases, to be used in subharmonic mixers for upconversion and downconversion. Therefore, it exploits a four-stage ring-oscillator. However, as shown in [21], an N-stage ring-oscillator can only provide accurate phases at its free running frequency, and the spacing of its output phases will deviate from $180/N$ when it is injection-locked to another frequency. This is a well-known issue and there are some solutions to maintain accurate output phase spacing across the locking range [21–23]. These solutions are mainly based on injecting the synchronization signal at multiple nodes using appropriate phases. These techniques will be studied in more detail in Chap. 4. As shown in [22], this phase error is directly proportional to the deviation from the free-running frequency. The frequency synthesizer of [18] uses digital calibration of the free-running frequency of the ring-oscillator to correct for this phase error. The details of this phase error correction are discussed in [22].

The injection strength needs to be large enough to achieve adequate phase noise suppression in the ring-oscillator and also to obtain a fast enough transient in the injection-locked oscillator. However, too large an injection signal will cause more systematic amplitude and phase error that can be beyond the level that can be corrected, while it will not provide any further phase noise suppression.

The synthesizer of [18] is implemented in a 90 nm CMOS technology and performs frequency hopping in approximately 4 nS. Using three separate PLLs to generate the synchronization signal at the LO frequency or half the LO frequency helps this frequency synthesizer to achieve a spurious tone level of better than −43 dBc, compared to −19 dBc of the frequency synthesizer of [17].

2.2.5 Delay-Locked Loop-Based Techniques

Another technique for frequency synthesis is based on combining the uniformly spaced phases of a reference signal obtained using a delay-locked loop (DLL) [24]. This technique achieves a frequency multiplication factor of N where N is the number of delay cells in the DLL. In this case, it is assumed that each delay cell introduces a phase shift of $2\pi/N$. This technique benefits from several advantages: first, in contrast to a PLL, a DLL is a first-order system; hence, it has no stability issue. Consequently, the bandwidth of a DLL can be chosen to be wide enough to meet the settling requirement. Second, the number of delay cells in the feedback loop can be programmable, hence a programmable frequency multiplication factor can be implemented. These two properties of a DLL-based frequency synthesizer make it a good candidate for fast agile applications, such as UWB. In addition to these properties, as discussed in [24], a DLL-based frequency synthesizer can achieve a better close-in phase noise when compared to a PLL-based frequency synthesizer. This comes from the fact that the power of the phase noise at the output of the delay cell is constant over the offset frequencies [25] while the phase noise power of a VCO is inversely proportional to the square of the offset frequency [26]. The superior close-in phase noise of a DLL-based frequency synthesizer may not be very important for UWB systems, but can be very important in many wireless communication standards.

A simplified block diagram of a DLL-based frequency synthesizer is shown in Fig. 2.15. As can be seen from this figure, the outputs of all the delay cells go to the edge combiner. The edge combiner combines all the N phases of the reference signal to generate a signal at N times the frequency of the reference signal. At low frequencies an edge combiner can be implemented using combinational logic gates including exclusive-OR (XOR) gates, similar to the clock multiplier presented in [27]. However, some analog techniques are required to implement an edge combiner at RF frequencies. A typical circuit implementation of such an edge combiner is shown in Fig. 2.16 [24]. As can be seen from Fig. 2.16, it consists of N open drain differential pairs that are all terminated to a tuned load.

The operation of a fifth-order DLL-based frequency synthesizer is studied in [28]. We present this analysis for the general case where the DLL consists of N stages.

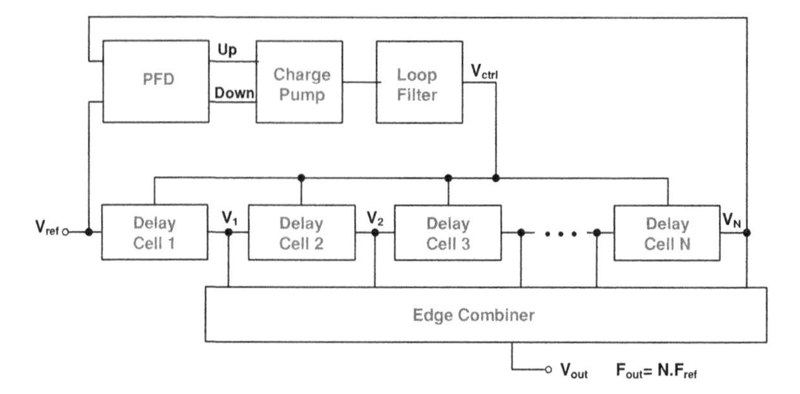

Fig. 2.15 A general block diagram of a DLL-based frequency synthesizer

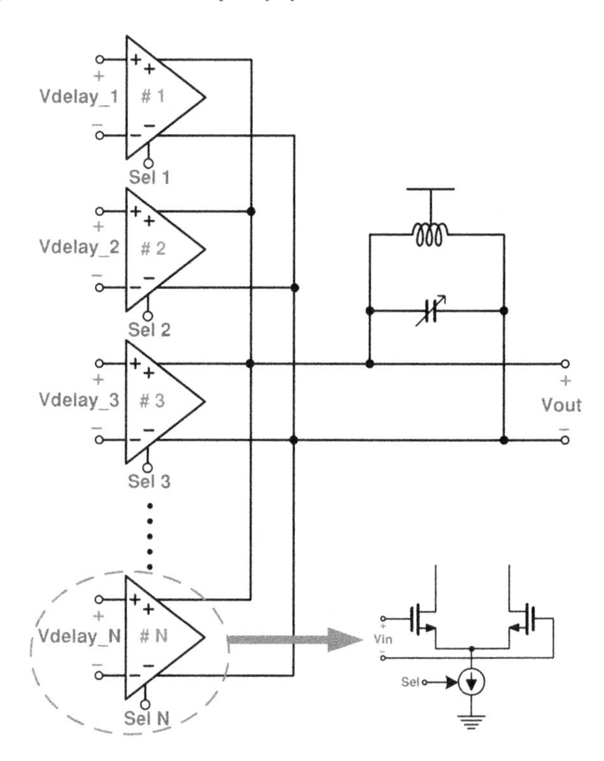

Fig. 2.16 An example of circuit implementation for edge combiner for a DLL-based frequency multiplier

To do so, consider the N-stage DLL of Fig. 2.15. We assume that the gain of each delay cell is unity, i.e., the amplitude at the output of all delay cells are equal which is approximately true in large signal operation. Furthermore, we assume that the delay of all the delay cells are well matched. If this DLL is locked to a reference signal with a fundamental period of T_i, each delay cell introduces a delay of T_D where

$$T_D = T_i/N. \tag{2.8}$$

If the edge combiner sums the outputs of the delay cells, the transfer function of the edge combiner can be expressed as shown below.

$$H(s) = \mathrm{e}^{-sT_D} + \mathrm{e}^{-s2T_D} + \mathrm{e}^{-s3T_D} + \cdots + \mathrm{e}^{-sNT_D} \tag{2.9}$$

If $s \neq j2\pi k/T_D$, where k in an integer number, (2.9) can be simplified as

$$H(s) = \mathrm{e}^{-sT_i/N} \frac{1 - \mathrm{e}^{-sT_i}}{1 - \mathrm{e}^{-sT_i/N}}. \tag{2.10}$$

Summing the inputs to the delay cells instead of the outputs of the delay cells in the edge combiner only changes the phase of the transfer function of (2.10) while it keeps the magnitude of (2.10) intact. The magnitude of the transfer function of (2.10) is evaluated by substituting s with $j\omega$ where $\omega = 2k\pi/T_i$ $(k = 1, 2, \ldots, N)$, as shown below.

$$|H(j\omega)| = 0 \quad \text{if } \omega = \frac{2\pi k}{T_i} \quad (k = 1, 2, \ldots, N - 1) \tag{2.11a}$$

$$|H(j\omega)| = N \quad \text{if } \omega = 0, \quad \frac{2\pi N}{T_i} \tag{2.11b}$$

As can be seen from (2.11a) and (2.11b), the transfer function of (2.10) has nulls at all the harmonics of $2\pi/T_i$ except at the Nth harmonic. Hence, it keeps the Nth harmonic of the reference signal at the output of the edge combiner and serves as a frequency multiplier by a factor of N. The zeros in the magnitude of the transfer function of (2.10) are obtained when the delay cells are well matched, and presence of any mismatch between the delays cells leads to spurious tones at the output spectrum, as discussed in [24].

The frequency multiplication factor of a DLL-based frequency synthesizer is the number of its delay cells, i.e., the frequency multiplication factor of DLL-based synthesizer can be changed by changing the number of delay cells that are used in the feedback loop, or changing the delay per cell provided that each delay cell has enough programmability range to cover the required range of delay per cell at the reference frequency.

The UWB frequency synthesizer of [25] is a DLL-based frequency synthesizer (multiplier) that generates the center frequencies for Band Group One and achieves a band switching time of less than 9.5 nS with a phase noise of −120 dBc/Hz at 1 MHz offset. The block diagram of this frequency synthesizer is shown in Fig. 2.17.

As discussed earlier, the bandwidth of a DLL can be chosen wide enough with no concern regarding the stability. However, the mathematical model of a DLL that is derived in [28] is only valid when the bandwidth of the DLL is only a fraction of the reference frequency of the DLL (the bandwidth is usually one tenth of the reference frequency) [28]. Consequently, a high reference frequency is needed to meet the

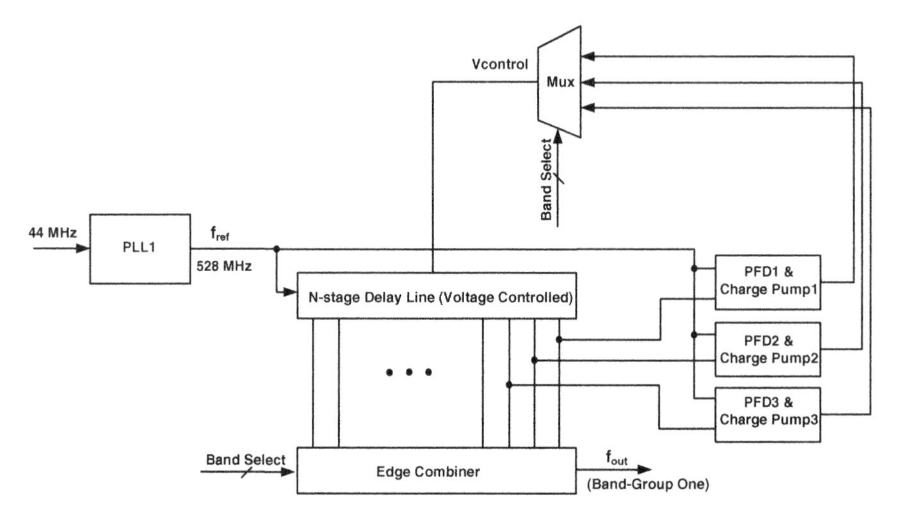

Fig. 2.17 The block diagram of the DLL-based frequency synthesizer of [25] for Band Group One of UWB spectrum

settling requirement for a UWB frequency synthesizer. The frequency synthesizer of [25] uses a reference frequency of 528 MHz along with multiplication factors of 13, 15, and 17 and is followed by a divide-by-two to generate the quadrature outputs for the center frequencies of Band Group One.

As discussed in [25], switching between the delay cells, to achieve band switching, can introduce a "glitch" on the feedback signal, which will cause a longer settling time than required for UWB system. This issue is solved in the frequency synthesizer of [25] by using multiple PFDs and charge-pumps, as shown in Fig. 2.17.

2.2.6 Comb Generator Technique

Another technique that is used in fast hopping frequency synthesizers is comb generator [29]. In this technique, a "comb" of frequencies is generated and is present at all times. Therefore, the speed of frequency hopping is limited to the speed of switching. Combination of filtering and frequency nulling maybe required to suppress the unwanted frequencies.

To explain the operation of a comb generator, we consider an ideal comb of frequencies that contains all the harmonics of a fundamental frequency f_o with equal amplitudes of A_o/T_o. The output of this comb generator in frequency domain is represented as follows.

$$S(f) = \frac{A_o}{T_o} \sum_{k=-\infty}^{+\infty} \delta(f - kf_o) \tag{2.12}$$

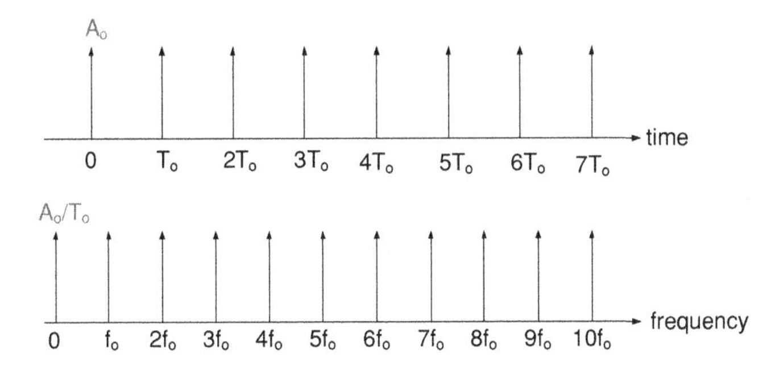

Fig. 2.18 Representation of the output of an ideal comb generator with a fundamental frequency of $f_o = 1/T_o$ in time and frequency domains

It can be concluded from (2.12) that the output of the comb generator is an impulse train in time domain too, i.e.,

$$s(t) = A \sum_{n=-\infty}^{+\infty} \delta(t - nT_o) \tag{2.13}$$

where $T_o = 1/f_o$. The output of an ideal comb generator with a fundamental frequency of f_o, or a fundamental period of T_o, in time and frequency domains is shown in Fig. 2.18.

It can be concluded from (2.13) that any circuit that can generate an impulse train at a given fundamental frequency of f_o can be used as a comb generator. To get one step closer to actual implementation of a comb generator, we assume that the impulse train in the time domain representation of the comb generator output shown in (2.13) is replaced by a pulse train with an amplitude of A_o, a pulse width of τ and a period of T_o. In this case, a Fourier series representation can be used to demonstrate the output of this new comb generator in time domain as follows.

$$s(t) = \frac{A_o \tau}{T_o} \sum_{n=-\infty}^{+\infty} \mathrm{sinc}(n\tau/T_o) e^{-j2\pi nt/T_o} \tag{2.14}$$

As can be seen from (2.14), in a realistic comb generator the sinc(·) function leads to non-equal amplitudes for the harmonics of the fundamental frequency. In addition, the limited bandwidth of the circuitry that implements the comb generator also leads to non-flat spectrum over frequency. On the other hand, the sinc(·) function in (2.14) can be used to null some of the undesired frequency components. The magnitude of the harmonics of the output of a comb generator that is described by (2.14) is shown in Fig. 2.19 for different values of τ/T_o. As can be seen from (2.14) and Fig. 2.19 a narrow pulse width ($\tau/T_o \ll 1$) is required to achieve a wideband comb generator.

As mentioned earlier, a wideband comb generator requires generating very narrow pulses. Step recovery diodes [30] are often used to generate very narrow pulses or

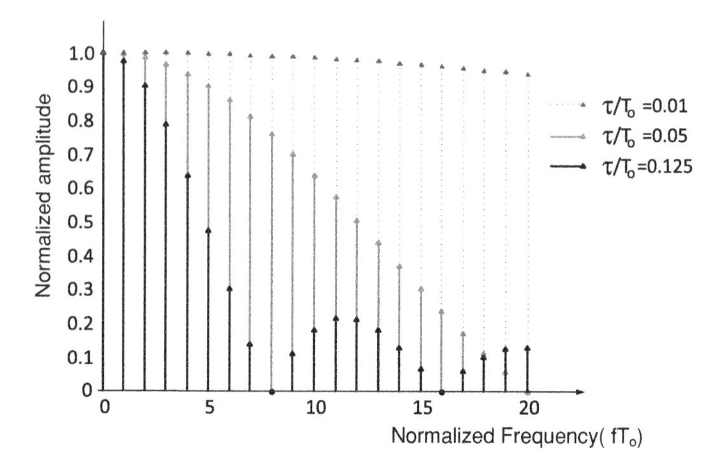

Fig. 2.19 Spectrum of the output of a comb generator with a fundamental frequency of $f_0 = 1/T_0$ and a pulse width of τ for different values of τ/T_0

to approximate an impulse. A hybrid approach is also shown in [31] to generate a comb of frequencies for optical communications. However, other techniques needs to be deployed for CMOS implementation of a comb generator.

A comb generator for UWB needs to generate integer multiples of 264 MHz. This fundamental frequency is used since it is the common divisor of all the UWB band center frequencies. In addition, it needs to provide a flat frequency response over the desired UWB frequency range. However, one of the main obstacles in implementing an integrated comb generator-based frequency synthesizer for UWB is the very high spurious level of the comb generator and the difficulty of on-chip implementation of a filter that provides adequate suppression of the spurious tones.

2.2.7 Techniques Based on Polyphase Filtering

The work presented in [32] uses a different technique to mitigate the agility requirement of a UWB frequency synthesizer. In the receiver of [32] a fixed frequency PLL (7.128 GHz) is used along with a fast switching polyphase filter and a high-speed Analog-to-Digital Converter (ADC).

The synthesizer used in the transceiver of [32] covers Band Group Three, and uses a fixed frequency PLL to generate the center frequency of band eight (7.128 GHz). The receiver of [32] uses quadrature phases of 7.128 GHz to downconvert the entire Band Group Three to DC. As a result, band eight is centered at DC, and bands seven and nine are centered at −756 and +756 MHz, respectively. The fast-switching polyphase filter of Fig. 2.20a is used to select a band of interest. The principles of operation of this polyphase filter is shown in Fig. 2.20b. As can be seen from Fig. 2.20b, the polyphase filter can pass the negative frequencies when

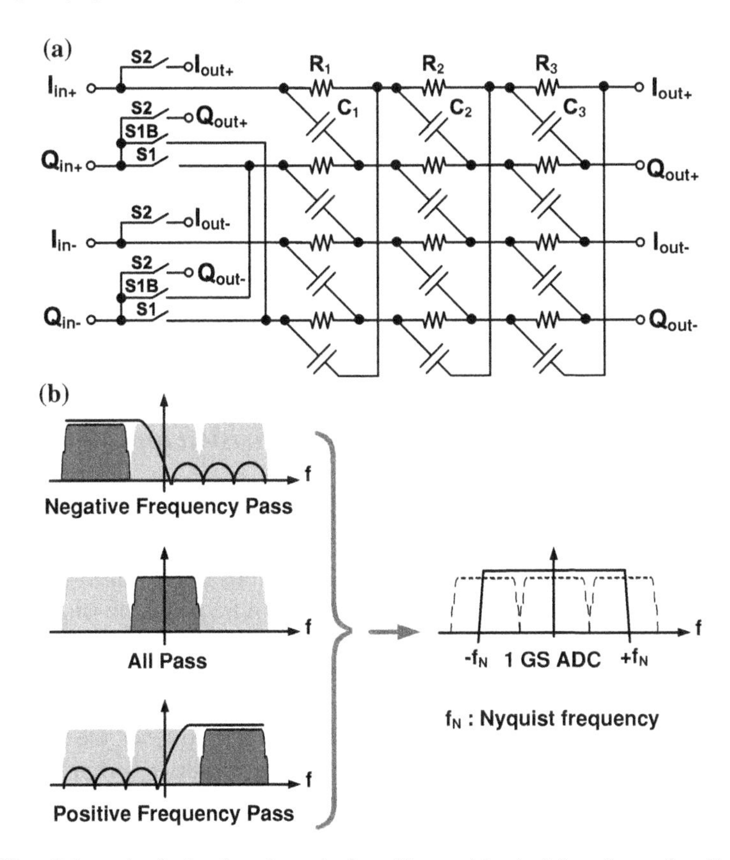

Fig. 2.20 **a** Schematic of a fast-hopping polyphase filter, and **b** principles of operation. By feeding the appropriate quadrature sequence, the polyphase filter can pass the positive or negative frequency band

operating in band seven, or positive frequencies when operating in band nine by using the appropriate quadrature sequence (clockwise or counterclockwise) at its input. The polyphase filter can also be bypassed for operation in band eight. The polyphase filter of Fig. 2.20a achieves a switching time of 5 ns. The polyphase filter is followed by a 1 GS/sec ADC. Final downconversion to DC for bands seven and nine will be done after analog-to digital conversion in digital domain.

The synthesizer presented in [32] eliminates the use of SSB mixers in the synthesizer path and does not suffer from mixing spurious tones. However, due to limitations in the speed and dynamic range of the ADCs, it is very difficult for this architecture to cover more than one Band Group using a single fixed frequency PLL.

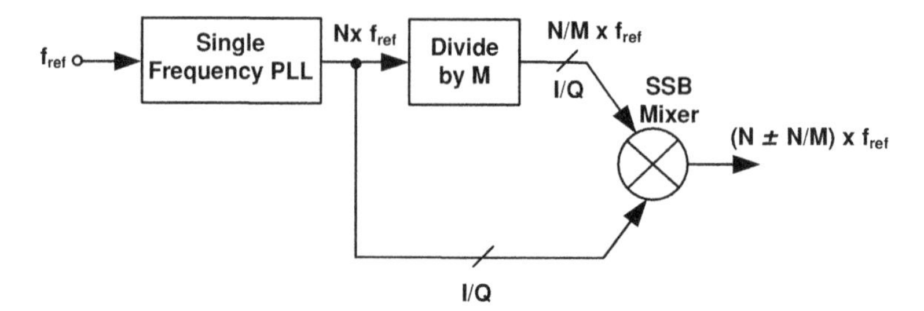

Fig. 2.21 Block diagram of a frequency synthesizers for UWB based on frequency division and mixing

2.2.8 Frequency Synthesis by Frequency Division and Mixing

Another architecture that can be used to implement a frequency synthesizer is a hybrid architecture shown in Fig. 2.21 where a combination of fixed frequency PLLs, frequency dividers, and SSB mixers are required. In this architecture, all the MB-OFDM UWB carriers are generated by combination of frequency division and SSB mixing. Accordingly, this technique alleviates the challenging settling requirement of the PLL.

The architecture of Fig. 2.21 can achieve a very fast switching time, as low as 1 nS. In addition, it can cover the entire UWB frequency range. However, the main drawback of this architecture is the use of SSB mixers. As was discussed in Sect. 2.2.1, using SSB mixers requires the availability of quadrature phases of both mixing signals, with sufficient amplitude and phase accuracy to achieve adequate side-band cancellation. On the other hand, every mixing stage introduces multiple spurious tones. Mixer linearization, and appropriate frequency planning can minimize spurious responses. However, mixer linearization reduces the conversion gain, and it will require additional power consumption to achieve same voltage swing at the mixer output.

A generalization to the architecture of Fig. 2.21 using two SSB mixers is shown in Fig. 2.22. The main challenge in implementing the architecture of Fig. 2.22 is to achieve a hardware efficient design compatible with digital CMOS technology.

For instance, the synthesizer of [33] requires two PLLs and one level of SSB mixing to generate the center frequencies of Band Group One. Another example is the frequency synthesizer of [34] which uses two separate PLLs and one SSB mixer to generate seven band center frequencies of MB-OFDM UWB (based on the earlier band allocation [14]).

The architecture of Fig. 2.22 is used in the majority of UWB frequency synthesizers that cover more than one Band Group, and in most of the 14-band UWB frequency synthesizers due to flexibilities in covering a wide frequency range and overcoming the settling issues [35–37].

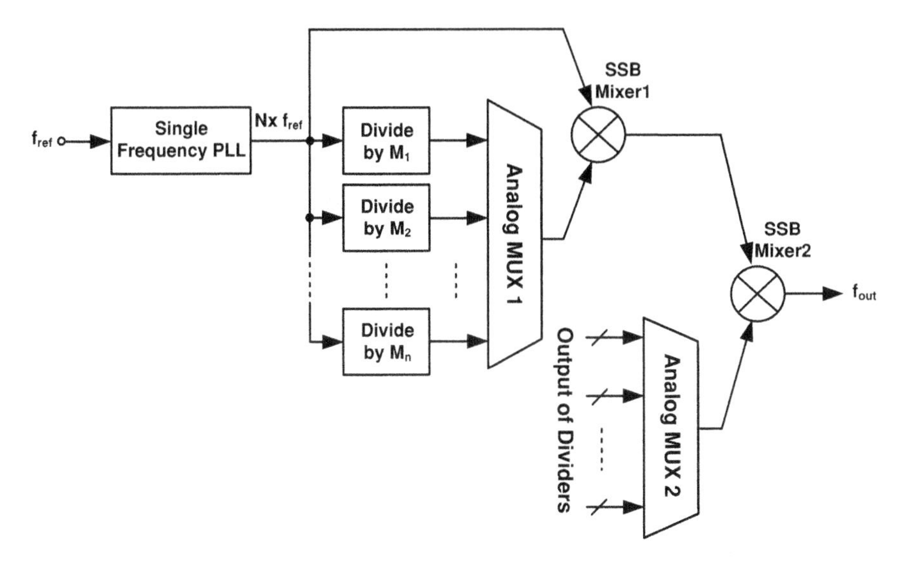

Fig. 2.22 Generalized architecture for a UWB frequency synthesizers that is based on frequency division and mixing with two levels of SSB mixing

2.2.9 Comparison of Different Frequency Synthesizer Solutions

Among all the different architectures for a UWB synthesizer, the all digital PLL presented in Sect. 2.2.3 and the hybrid architecture based on frequency division and mixing presented in Sect. 2.2.8 seem to meet the agility requirement of UWB, and at the same time cover the entire UWB span with a manageable power consumption and die area. The use of an AD- PLL-based architecture for a UWB frequency synthesizer eliminates the use of SSB mixers and frequency dividers required for frequency mixing. Hence, there will be no concern with mixer nonlinearity. However, the objective of this work was to explore the inductor-less implementation of frequency division and mixing techniques in a CMOS-only technology. In addition, although it is not required by WiMedia v1.5 [16, 38], but our focus in choosing an architecture for a UWB frequency synthesizer was to cover the entire fourteen-band span of UWB. In addition, the ability to perform frequency hopping among different Band Groups is also considered. The architecture that is proposed in Sect. 2.3 meets all these specifications.

2.3 Proposed Fourteen-Band Frequency Synthesizer for UWB

The latest Band Groups and band allocation for MB-OFDM UWB was formerly shown in Fig. 1.9. Based on that, a frequency plan that can generate the band center frequencies for all fourteen bands and perform frequency hopping within each of the six Band Groups is shown in Fig. 2.23.

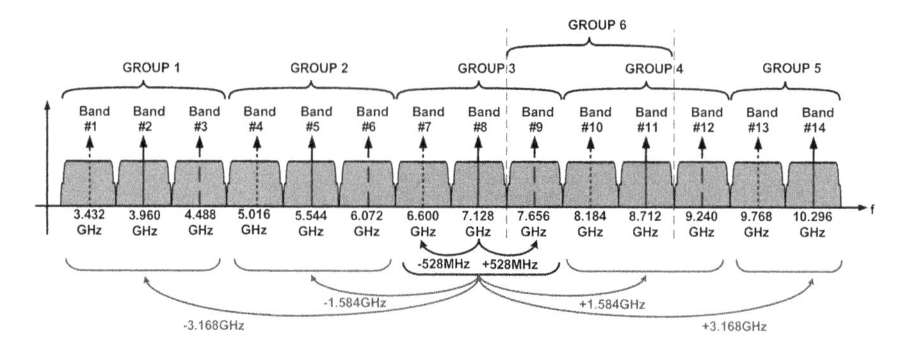

Fig. 2.23 Proposed UWB synthesizer frequency plan

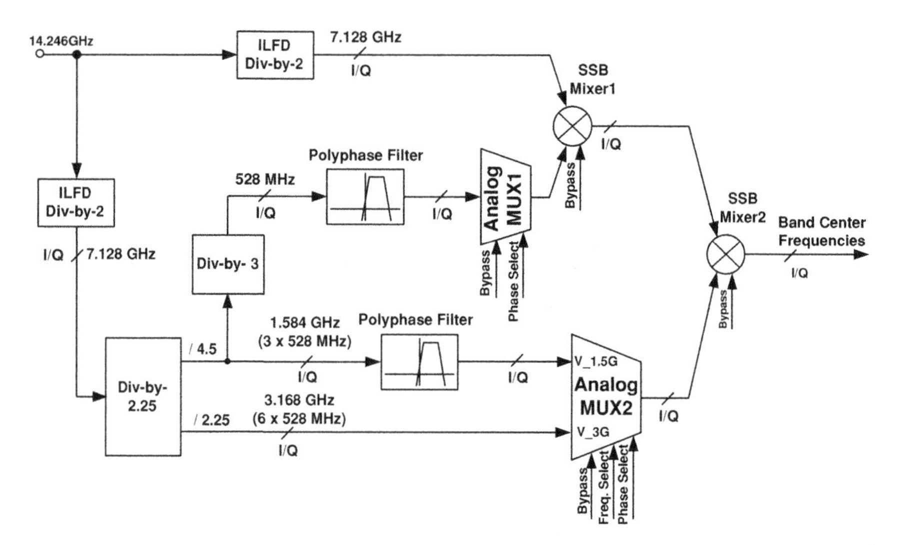

Fig. 2.24 Architecture for the universal fourteen-band UWB frequency synthesizer

As described by (1.2) each center frequency for the bands shown in Fig. 2.23 can be expressed by $f_c = (5.5+n) \times 528$ MHz where n is the band number from 1 to 14. This implies that the band center frequencies are not integer multiples of the channel spacing, and some sort of fractional frequency division ratio needs to be employed to derive the channel spacing (528 MHz) from any of the band center frequencies.

Moreover, the frequency plan of Fig. 2.23 requires quadrature phases of 1.584 and 3.168 GHz (three and six times the channel spacing respectively). For this purpose, the center frequency of Band 8 is very useful, since all the required translation frequencies (528 MHz, 1.584 and 3.168 GHz) can be derived from it when the fractional frequency dividers of Fig. 2.24 are used, in a similar manner to the three-band synthesizer of [39]. Figure 2.24 shows our proposed architecture to implement the frequency plan shown in Fig. 2.23.

The proposed frequency synthesizer uses a single external 14.256 GHz input. Quadrature signals for the operation of the first mixing stage are generated from a ring-oscillator-based Injection Locked Frequency Divider (ILFD) whose output is the center frequency of Band 8. The channel spacing of 528 MHz is obtained from the output of two additional regenerative dividers, with total division ratio of 13.5. A cascade of quadrature output divide-by-2.25, divide-by-2, and divide-by-3 circuits provides the required frequencies. As shown in Fig. 2.24, two similar frequency dividers are used to generate quadrature phases at 7.128 GHz. This scheme avoids any I/Q amplitude or phase mismatch induced due to routing, compared to the case when only one frequency divider is used.

2.3.1 Frequency Dividers with Quadrature Outputs

As can be seen from Fig. 2.24 the inductor-less implementation of this technology relies on using SSB mixers and the availability of quadrature phases of all the signals used in the mixers. Therefore, frequency dividers that generate quadrature outputs and can be implemented in digital CMOS technology while running at several GHz frequencies are of interest. Ring-oscillator-based frequency dividers that can meet these goals are studied in Chap. 3.

2.3.2 Fractional Frequency Dividers

The architecture of Fig. 2.24 requires some frequency dividers that can obtain fractional division ratios while providing quadrature outputs. Several techniques to implement a fractional division ratio are studied in Chap. 4 and the appropriate architecture that can be implemented in digital CMOS is chosen. In addition, the operation and the stability of this divider is studied.

2.3.3 Broadband High-Gain Interface Amplifiers

The UWB frequency synthesizer of Fig. 2.24 uses up to two levels of SSB mixing to generate the desired carrier frequency. The output of the first SSB mixer needs to be amplified to be used in the second SSB mixer. Different amplifiers that can be used for this purpose in the frequency synthesizer of Fig. 2.24 are studied in Chap. 5 and an architecture that can be implemented in a low-voltage digital CMOS technology is chosen.

References

1. Uusikartano R, Niittylahti J (2001) A periodical frequency synthesizer for a 2.4-GHz fast frequency hopping transceiver. IEEE Trans Circuits Syst II: Analog Digit Signal Process 48(10):912–918
2. Kroupa V (1999) Direct digital frequency synthesizers. IEEE Press, New York
3. Sandner C, Derksen S, Draxelmayr D, Ek S, Filimon V, Leach G, Marsili S, Matveev D, Mertens K, Michl F et al (2006) A WiMedia/MBOA-compliant CMOS RF transceiver for UWB. IEEE J Solid-State Circuits 41(12):2787–2794
4. Perrott MH (1997) Techniques for high data rate modulation and low power operation of fractional-N frequency synthesizers. Ph.D. dissertation, Massachusetts Institute of Technology
5. Gardner F (2005) Phaselock techniques. Wiley-Blackwell, New York
6. Gardner F (1980) Charge-pump phase-lock loops. IEEE Trans Commun 28(11):1849–1858
7. Bishap R, Dorf R (2004) Modern control systems. Addison wesley publishing company, Reading
8. Rategh H, Lee T (2001) Multi-GHz frequency synthesis and division: frequency synthesizer design for 5 GHz wireless LAN systems. Springer
9. Tak G, Hyun S, Kang T, Choi B, Park S (2005) A 6.3–9-GHz CMOS fast settling PLL for MB-OFDM UWB applications. IEEE J Solid-State Circuits 40(8):1671–1679
10. Vaucher C (2000) An adaptive PLL tuning system architecture combining high spectral purity and fast settling time. IEEE J Solid-State Circuits 35(4):490–502
11. Razavi B, Aytur T, Lam C, Yang F, Li K, Yan R, Kang H, Hsu C, Lee C (2005) A UWB CMOS transceiver. IEEE J Solid-State Circuits 40(12):2555–2562
12. Bergervoet J, Harish K, Lee S, Leenaerts D, van de Beek R, van der Weide G, Roovers R (2007) A WiMedia-compliant UWB transceiver in 65 nm CMOS. IEEE ISSCC digest of technical papers, pp 112–590, Feb 2007
13. Stadius K, Rapinoja T, Kaukovuori J, Ryynanen J, Halonen K (2007) Multitone fast frequency-hopping synthesizer for UWB radio. IEEE Trans Microw Theory Tech 55(8):1633–1641
14. Multi-Band OFDM Physical Layer Proposal, July 2003, doc IEEE 802.15- 03/267r5
15. Leenaerts D, van de Beek R, Bergervoet J, Kundur H, van der Weide G, Kapoor A, Pu T, Fang Y, Wang Y, Mukkada B et al (2009) A 65 nm CMOS inductorless triple Band Group WiMedia UWB PHY. IEEE J Solid-State Circuits 44(12):3499–3510
16. Standard ECMA-368: High Rate Ultra Wideband PHY and MAC Standard, 3rd edition, Dec. 2008. http://www.ecma-international.org/publications/standards/Ecma-368.htm
17. Dal Toso S, Bevilacqua A, Tiebout M, Marsili S, Sandner C, Gerosa A, Neviani A (2008) UWB fast-hopping frequency generation based on sub-harmonic injection locking. IEEE J Solid-State Circuits 43(12):2844–2852
18. Mazzanti A, Vahidfar M, Sosio M, Svelto F (2009) A reconfigurable demodulator with 3-to-5GHz agile synthesizer for 9-band WiMedia UWB in 65nm CMOS. IEEE ISSCC digest of technical papers, pp 412–413, 413a, Feb 2009
19. Verma S, Rategh H, Lee T (2003) A unified model for injection-locked frequency dividers. IEEE J Solid-State Circuits 38(6):1015–1027
20. Dal Toso S, Bevilacqua A, Tiebout M, Marsili S, Sandner C, Gerosa A, Neviani A (Dec. 2008) UWB fast-hopping frequency generation based on sub-harmonic injection locking. IEEE J Solid-State Circuits 43(12):2844–2852
21. Chien J, Lu L (2007) Analysis and design of wideband injection-locked ring-oscillators with multiple-input injection. IEEE J Solid-State Circuits 42(9):1906–1915
22. Mazzanti A, Vahidfar M, Sosio M, Svelto F (2010) A low phase-noise multiphase LO generator for wideband demodulators based on reconfigurable subharmonic mixers. IEEE J Solid-State Circuits 45(10):2104–2115
23. Farazian M, Gudem P, Larson L (2010) Stability and operation of injectionlocked regenerative frequency dividers. IEEE Trans Circuits Syst I: Regul Pap 57(8):2006–2019

24. Chien G, Gray P (2000) A 900-MHz local oscillator using a DLL-based frequency multiplier technique for PCS applications. IEEE J Solid-State Circuits 35(12):1996–1999
25. Lee, T-C, Hsiao K-J (2006) The design and analysis of a DLL-based frequency synthesizer for UWB application. IEEE J Solid-State Circuits 41(6):1245–1252
26. Hajimiri A, Lee T (1998) A general theory of phase noise in electrical oscillators. IEEE J Solid-State Circuits 33(2):179–194
27. Chin-Chieh C, Chao-Ping S, Yen-Kuang C (2005) Clock multiplier. US Patent US 6977536 B2, 20 Dec 2005
28. Chien G (2000) Low-noise local oscillator design techniques using a DLL-based frequency multiplier for wireless applications. Ph.D. dissertation, University of California, Berkeley
29. Adkins L (1976) Fast frequency hopping with Surface Acoustic Wave (SAW) frequency synthesizers. Proceedings of the 30th annual symposium on frequency control, pp 367–370
30. Moll J, Hamilton S (1969) Physical modeling of the step recovery diode for pulse and harmonic generation circuits. Proc IEEE 57(7):1250–1259
31. Tsao S, Wu J, Chen T, Wu J (2002) Phaselocked tunable subcarrier comb generator. Electron Lett 30(24):2059–2060
32. Tanaka A, Numata K, Kodama H, Ishikawa H, Oshima N, Yano H (2009) A 2.88Gb/s digital hopping UWB transceiver. IEEE ISSCC digest of technicsl papers, pp 318–319, Feb 2009
33. Leenaerts D, van de Beek R, Van der Weide G, Bergervoet J, Harish K, Waite H, Zhang Y, Razzell C, Roovers R, Res P et al (2005) A SiGe BiCMOS 1ns fast hopping frequency synthesizer for UWB radio. IEEE ISSCC digest of technical papers, pp 202–593, Feb 2005
34. Lee J, Chiu D (2005) A 7-band 3–8 GHz frequency synthesizer with 1 ns bandswitching time in 0.18 μm CMOS technology. IEEE ISSCC digest of technical papers, pp 204–593, Feb 2005
35. Ismail A, Abidi A (2005) A 3.1- to 8.2-GHz zero-IF receiver and direct frequency synthesizer in 0.18μm SiGe BiCMOS for mode-2 MB-OFDM UWB communication. IEEE J of Solid-State Circuits 40(12):2573–2582
36. Liang C-F, Liu S-I, Chen Y-H, Yang T-Y, Ma G-K (2006) A 14-band frequency synthesizer for MB-OFDM UWB application. IEEE ISSCC digest of technical papers, pp 428–437, Feb 2006
37. Lu T-Y, Chen W-Z (2008) A 3-to-10GHz 14-band CMOS frequency synthesizer with spurs reduction for MB-OFDM UWB system. IEEE ISSCC digest of technical papers, pp 126–601, Feb 2008
38. Multiband OFDM physical layer specification PHY Specification: Final Deliverable 1.5 (2009). http://www.wimedia.org/en/docs/10003r02WM_CRB-WiMedia_PHY_Spec_1.5.pdf
39. Lin C, Wang C-K (2005) A regenerative semi-dynamic frequency divider for mode-1 MB-OFDM UWB hopping carrier generation. IEEE ISSCC digest of technical papers, vol 1. pp 206–207, Feb 2005

Chapter 3
Frequency Division and Quadrature Signal Generation at Microwave Frequencies

3.1 Overview

In many receiver and transmitter architecture, or in general any application that requires frequency translation, the availability of multiple phases of a carrier frequency is desired and makes the on-chip implementation of that architecture feasible or simpler. An example is an image-reject receiver using the Hartley or Weaver architectures, where quadrature phases of the carrier signal are needed [1]. There are several techniques to generate multiple phases of a carrier signal. One direct way is by using a ring-oscillator, or any multi-phase VCO, in the phase-locked loop. Other techniques include using a polyphase filter or a delay-locked loop to generate multiple phases of a carrier. In all these techniques, different phases of the carrier signal are generated at the carrier frequency. Another method to generate multiple phases of a carrier signal is based on synthesizing a carrier signal with a frequency that is a multiple of the frequency of the desired carrier, and then using frequency dividers to generate multiple phases of the desired signal after frequency division. This technique is very useful when we are interested in quadrature output phases, since many dividers, usually divide-by-two circuits, can provide quadrature output phases. In addition, as discussed in [1] most image-reject transceivers require quadrature phases of the carrier signal.

Quadrature generation with a frequency divider (a divide-by-two or a cascade of divide-by-two circuits) is the preferred technique in most wireless communication applications, especially at higher frequencies where stringent phase noise specifications prevent the use of ring-oscillators, and the quadrature accuracy requirements are too stringent to use polyphase filtering for quadrature generation. On the other hand, this technique for quadrature generation can be combined with frequency synthesis in phase-locked loops (PLLs), when the desired multi-phase frequency divider is used in the feedback path of a phase-locked loop. For these reasons, we put the focus of this chapter on the implementation of frequency dividers that can operate at high frequencies and provide multiple output phases (quadrature phases in particular for the purpose of this book).

M. Farazian et al., *Fast Hopping Frequency Generation in Digital CMOS*,

DOI: 10.1007/978-1-4614-0490-3_3,

High-speed frequency dividers are also key building blocks in the implementation of high-frequency PLLs. There have been many efforts to implement low power, area efficient, frequency dividers in the 60 GHz region [2, 3]. Implementing dividers with division ratios of larger than two can ease frequency synthesis at high frequencies and reduce power consumption and die area. Static frequency dividers work well up to a fraction of the transition frequency (f_T) of a device technology and beyond that limit their power consumption becomes extremely high. Moreover, they require a large signal swing, which is not easy to achieve at frequencies close to the f_T of a device technology. In addition, static frequency dividers usually achieve a division ratio of two, and to achieve larger division ratios, a cascade is required.

Injection-locked frequency dividers (ILFD) can work at higher frequencies compared to static frequency dividers. However, they usually suffer from a narrow input frequency locking range. Several groups have reported regenerative or ILFD working at frequencies up to 70 GHz [2, 4]. However, most of these designs cannot supply quadrature phases at the output. Furthermore, these architectures are inductor based, which may require a large die area. In addition, having tuned circuits implies a narrow range of operation.

The goal of this work is to implement multi-phase frequency dividers capable of operating at frequencies close to f_T with division ratios of larger than two. For compatibility with digital CMOS technology, the frequency divider must be able to operate at supply voltages as low as 1.2 V, and an inductor-less design methodology is adopted which leads to a smaller die area. But the power consumption of such an approach may be higher in the absence of tuned circuits and inductors.

3.2 Injection-Locked Frequency Divider Design

As was mentioned earlier, the goal of this work is to achieve both frequency division and multiple phases of the output at microwave frequencies. Because of its practical importance, our focus here will be mainly on achieving quadrature outputs at the output of frequency dividers.

There are certain techniques to generate quadrature output phases, including filter-based techniques, e.g., using a high-order polyphase filters to generate quadrature outputs, and ring-oscillator-based techniques.

Polyphase filter-based techniques to generate quadrature phases of signal are usually implemented using a cascade of a frequency divider and a polyphase filter. As a result, the frequency division and quadrature generation are not performed at the same time. One limitation of this method is the insertion loss of on-chip filters that can lead to significant loss of signal. Consequently, post-filtering amplification is required which adds to the power consumption and the die area.

On the other hand, quadrature signal generation techniques that use a ring-oscillator-based frequency divider can obtain multiple phases of the output and frequency division at the same time. Proper choice of the number of stages can provide the desired output phases. An N-stage ring-oscillator provides a phase shift

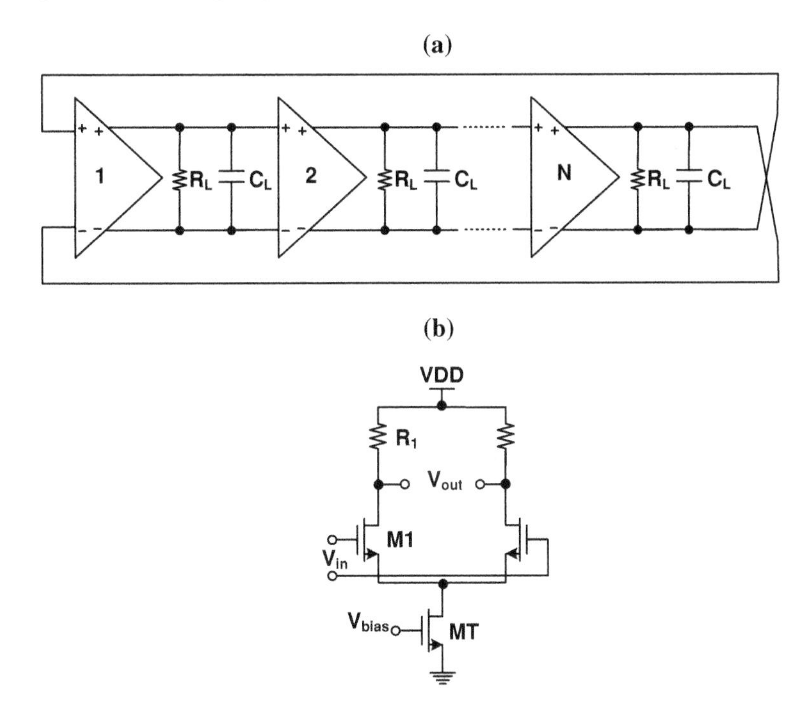

Fig. 3.1 **a** *N*-stage ring-oscillator, **b** resistive load differential pair delay cell

of π/N per stage at its self-resonance oscillation frequency. As a result, a two-stage or four-stage ring-oscillator is required to obtain quadrature output phases.

The Barkhausen criteria for the loop gain of the oscillators [5] puts a minimum required gain per stage in any ring-oscillator, and this requirement is more relaxed for a four-stage oscillator than a two-stage ring-oscillator. This becomes more important while moving to higher frequencies, and it would justify using four-stage ring-oscillators. Therefore, in this section, we concentrate on ring-oscillators with more than two stages, to combine the functions of frequency division and quadrature output generation. Two-stage ring-oscillators are studied in detail in Sect. 4.5.

Consider an *N*-stage ring-oscillator (Fig. 3.1a). As shown in [6], the oscillation frequency of the ring-oscillator is given by

$$f_{\mathrm{osc}} \approx \frac{1}{2N R_{\mathrm{L}} C_{\mathrm{L}} \ln 2} \tag{3.1}$$

where R_{L} and C_{L} are respectively the equivalent resistance and capacitance at the output of each delay cell.

At the oscillation frequency each stage must introduce a phase shift of π/N to satisfy the criteria for oscillation. This ring-oscillator can be implemented using the differential pair delay stage with resistive load shown in Fig. 3.1b. However, more than two delay stages are required to meet the phase shift requirement.

Here, we generate quadrature phases at the output, so a ring-oscillator is required with at least four delay stages, when the delay stage in Fig. 3.1b is used. Increasing the number of stages beyond four increases the area and power dissipation, and reduces the achievable self-resonance frequency (SRF).

To analyze injection locking in ring-oscillators, we use the nonlinear ILFD model introduced in [7], which is shown in Fig. 3.2. The input signal is injected to the tail current source of the first delay stage of the ring-oscillator, which is modeled as a single-balanced mixer. The function $f[\cdot]$ models the nonlinearity caused by the differential pair in commutating the tail current. The nonlinearity of $f[\cdot]$ introduces harmonics of ω_o prior to mixing. The "N-stage LPF" in Fig. 3.2 is the equivalent low-pass filter formed by the output resistance (R_L) and capacitance (C_L) of each stage in the N-stage ring-oscillator of Fig. 3.1a. The current at the mixer output (the drain current of M1 in Fig. 3.1b) can be written as

$$I_D(t) = g_m A_{\text{inj}} \sin(\omega_{\text{inj}} t + \phi_{\text{inj}}) f(V_O) \tag{3.2}$$

where g_m is the transconductance of the tail current source (MT) in Fig. 3.1b, and V_O is the output of the ILFD as shown in Fig. 3.2. For simplicity, the LO-to-output leakage of the single balanced mixer is not considered in (3.2), since this term does not contribute to the locking range or division ratio of the ILFD [8].

Since V_O is a periodic signal, $f(V_O)$ can be expressed using a Fourier series expansion of the harmonics

$$f(V_O) = \sum_{k=-\infty}^{\infty} a_k e^{jk\omega_o t} \tag{3.3}$$

where a_k coefficients are the Fourier coefficients of the output of the ILFD of Fig. 3.2 (V_O). If V_O is large enough, has a 50 % duty cycle and the nonlinearity of the $f[\cdot]$ has odd symmetry, and we can estimate $f[\cdot]$ by a ± 1 square wave. In this case, the differential output current of the mixer can be expressed as

$$I_D(t) = \sum_{k=1}^{\infty} (\pm) \frac{2 g_m A_{\text{inj}}}{(2k-1)\pi} \cos[\omega_{\text{inj}} t + \phi_{\text{inj}} \mp (2k-1)\omega_o t] \tag{3.4}$$

The "N-stage LPF" shown in Fig. 3.2 removes the high-frequency mixing components. The mixing components must be at the SRF of the ILFD in order to sustain in the loop. In other words,

$$\left|\omega_{\text{inj}} - (2k-1)\omega_o\right| = \omega_o. \tag{3.5}$$

Assuming $\omega_{\text{inj}} > (2k-1)\omega_o$ (low side injection in the mixer), it can be concluded that

$$\frac{\omega_o}{\omega_{\text{inj}}} = \frac{1}{2k} \tag{3.6}$$

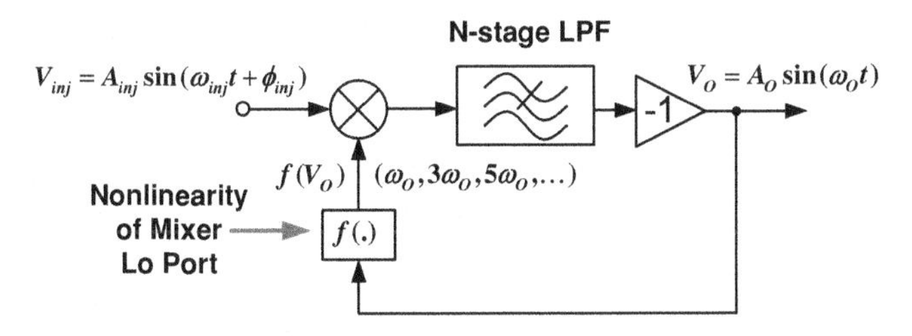

Fig. 3.2 Nonlinear model for ring-oscillator-based ILFD

The ILFD must be locked to the $2k$th harmonic of its SRF. In this case, ω_{inj} and the $(2k-1)$th harmonic of the SRF (ω_o) satisfy (3.5). On the other hand, the $(2k+1)$th harmonic of the SRF, which corresponds to high-side injection in the mixer, also satisfies (3.5). Therefore, after low-pass filtering, (3.4) can be simplified as follows.

$$I_D(t) \cong \frac{4g_{\mathrm{m}} A_{\mathrm{inj}}}{\pi(4k^2-1)} \left(2k \cos\phi_{\mathrm{inj}} \cos\omega_o t - \sin\phi_{\mathrm{inj}} \sin\omega_o t\right) \tag{3.7}$$

The upper limit of the mixer output current derived in (3.7) is, therefore,

$$\left|I_{D_{\max}}\right|_{\omega=\omega_o} < 4g_{\mathrm{m}} A_{\mathrm{inj}} \frac{2k}{\pi(4k^2-1)} \tag{3.8}$$

As can be seen in 3.7 and 3.8, the mixer output current drops inversely with the division ratio. This leads to a reduction of the input sensitivity of the ILFD when injection-locked to higher order harmonics of ω_o. This leads to the well-known narrower input frequency range for larger division ratios. In this work, we use a tuning mechanism to compensate for this problem.

3.3 Circuit Implementation

The quadrature output ring-oscillator-based ILFD is shown in Fig. 3.3. It consists of four delay stages. This divider generates eight different phases of the output signal. However, as can be seen from Fig. 3.3 only four phases (differential quadrature phases) are buffered and used since the focus of this work was quadrature generation.

Compared to two- or three-stage ring-oscillators, a four-stage ring-oscillator relaxes the gain requirement of each stage to meet the loop gain criteria. As a result, a smaller load resistor is used in the delay cell, which will allow the ring-oscillator to achieve a higher SRF, but this will increase the required power consumption to achieve the desired voltage swing. As was discussed in Sect. 3.2, the ILFD needs some tuning mechanism to overcome the narrow locking range problem. An additional tuning element will add some parasitic capacitance, and limit the maximum

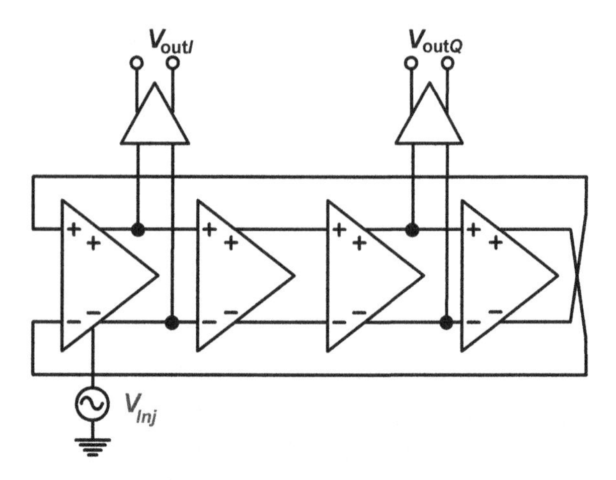

Fig. 3.3 Four-stage ring-oscillator-based injection-locked frequency divider (ILFD)

achievable SRF, so the SRF is tuned by changing the bias current of each delay cell. Changing the bias current directly affects the output impedance of each cell, which changes the SRF of the ring-oscillator, as expressed by (3.1) [8].

If we are only interested in differential quadrature output phases, the four-stage ring-oscillator of Fig. 3.3 can be modified to become the ring-oscillator of [9]. The four-stage ring-oscillator of [9] is shown in Fig. 3.4a. In this scheme, the first and the third stages are very similar to the resistive load differential pair delay cell of Fig. 3.1b, but the second and the fourth stages are implemented using a different delay cell with a tuning mechanism shown in Fig. 3.4c. In addition to the resistive delay cell of Fig. 3.4b, the delay cells that are used in the second and the fourth stages of the four-stage ring-oscillator of Fig. 3.4a consist of a cross-coupled NMOS pair. The differential V_{TUNE} applied to the current sources changes the negative resistance of this delay cell by changing the portion of the bias current that goes to the cross-coupled NMOS pair, hence it can change the delay of this delay cell. In short, this scheme provides a coarse tuning mechanism to change the total delay of the loop, and hence changes the SRF of the ring-oscillator of Fig. 3.4a. Consequently, it widens the input range of the ILFD that is implemented using this ring-oscillator. However, any mismatch between the delay cells used for the first and the third stages translates into I/Q phase mismatch.

Another modification to the ring-oscillator-based ILFD of Fig. 3.3 is to replace the load resistors in the delay cells with transistors biased in the triode region. This will add more parasitic capacitance to each delay cell, and can lower the SRF, but adds the ability to change the load resistance easily and hence adjust the free-running frequency of the ring-oscillator to cover a wide input frequency range, as shown in [10]. There are several ways to automatically tune the SRF of an ILFD. It becomes more important when an ILFD has a narrow input range or it is used in a phase-locked loop. The ring-oscillator-based divide-by-four ILFD presented in [11] uses a digital calibration algorithm to tune its SRF. The three-stage ring-oscillator-based ILFD of [12] is locked to the fourth harmonic of its SRF; hence, it achieves a division

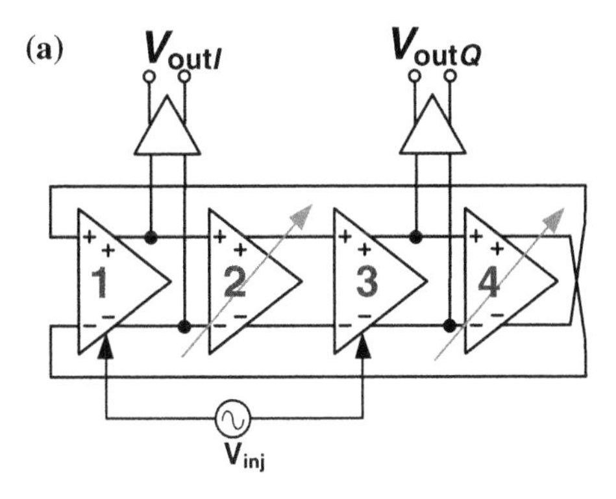

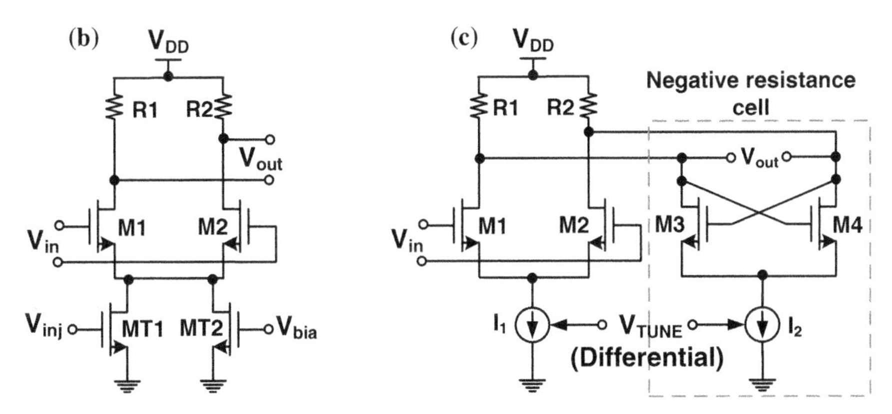

Fig. 3.4 **a** Schematic of the four-stage ring-oscillator-based ILFD of [9], **b** schematic of the resistive load differential pair delay cell used in stages one and three, **c** schematic of the tunable negative resistance delay cell used in stages two and four

ratio of four. This frequency divider also uses a 5-bit programmable resistor load (implemented using triode-biased transistors) along with a calibration algorithm to cover the input frequency range of 57–66 GHz.

3.4 Measurement Results of the Prototype Frequency Divider

The multi-phase ILFD of Fig. 3.3 is implemented in a 0.13 μm CMOS technology. Measured input sensitivity curves at different division ratios are plotted in Fig. 3.5. This ILFD achieves a locking range of roughly 5.5 GHz when operated as a divide-by-two, a locking range of 1.4 GHz when operated as a divide-by-four, and a locking range of 1 GHz for divide-by-six mode.

The measured time domain waveforms of the input and output of the frequency divider for different division ratios are shown in Fig. 3.6. In all these measurements,

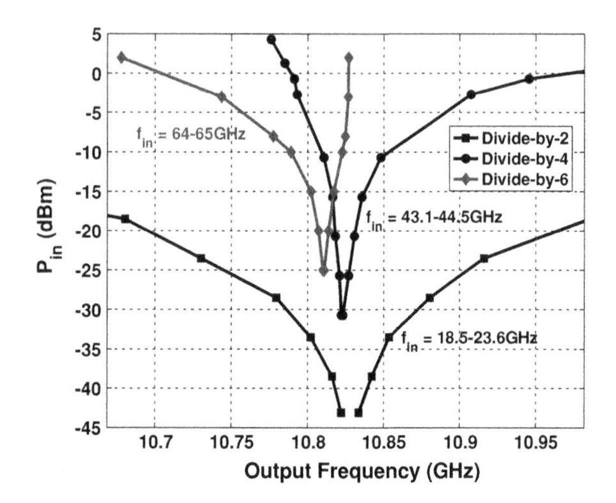

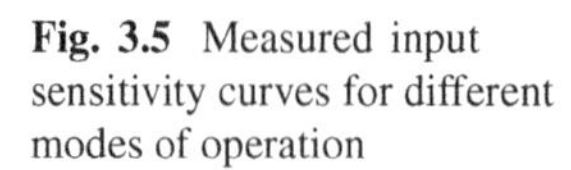
Fig. 3.5 Measured input sensitivity curves for different modes of operation

a single-ended input is applied to the ILFD. In Fig. 3.6a, the input is at 17.5 GHz and the ILFD achieves division ratio of two. Similarly in Fig. 3.6b and c division ratios of four and six are achieved for input signals at 38.8 and 64.6 GHz, respectively.

The differential quadrature phases of the output when the ILFD is operating as a divide-by-six are shown in Fig. 3.7. The I/Q phase and amplitude mismatch are roughly 4° and 0.5 dB for a 55 GHz input. As shown in Fig. 3.3, the ILFD is tested with a single-ended input applied to the first delay cell of the four-stage ring-oscillator. The I/Q phase and amplitude mismatch can be improved by injecting the input signal differentially. However, off-chip generation of complementary phases of a 55 GHz input is challenging and the ILFD was tested with a single phase input signal only. In addition, a phase and amplitude mismatch of 4° and 0.5 dB is acceptable for many applications.

Tuning curves of this ILFD for operation in the divide-by-six mode are shown in Fig. 3.8. Although these curves are plotted for 50 mV steps in VDD, this tuning can be done continuously. These curves show the possibility of extending the effective input range of the ILFD to 51–65 GHz. Moreover, this tuning capability provides sufficient margin to compensate for process variations.

The phase noise of the free-running ILFD and the phase noise of the ILFD under locked conditions for a 17.5 GHz input are plotted in Fig. 3.9. Despite the poor phase noise of ring-oscillators, they track the phase noise of the input source when they are injection-locked [13]. This ILFD achieves phase noise of −104 and −116 dBc/Hz at 10 kHz and 1 MHz offsets, respectively.

Figure 3.10 shows the measured SRF and the measured phase noise of the ILFD, when operated as a divide-by-six, for different values of external tuning. As is shown in this figure, the phase noise is better than −110 dBc/Hz at 1 MHz offset for all the values of external tuning.

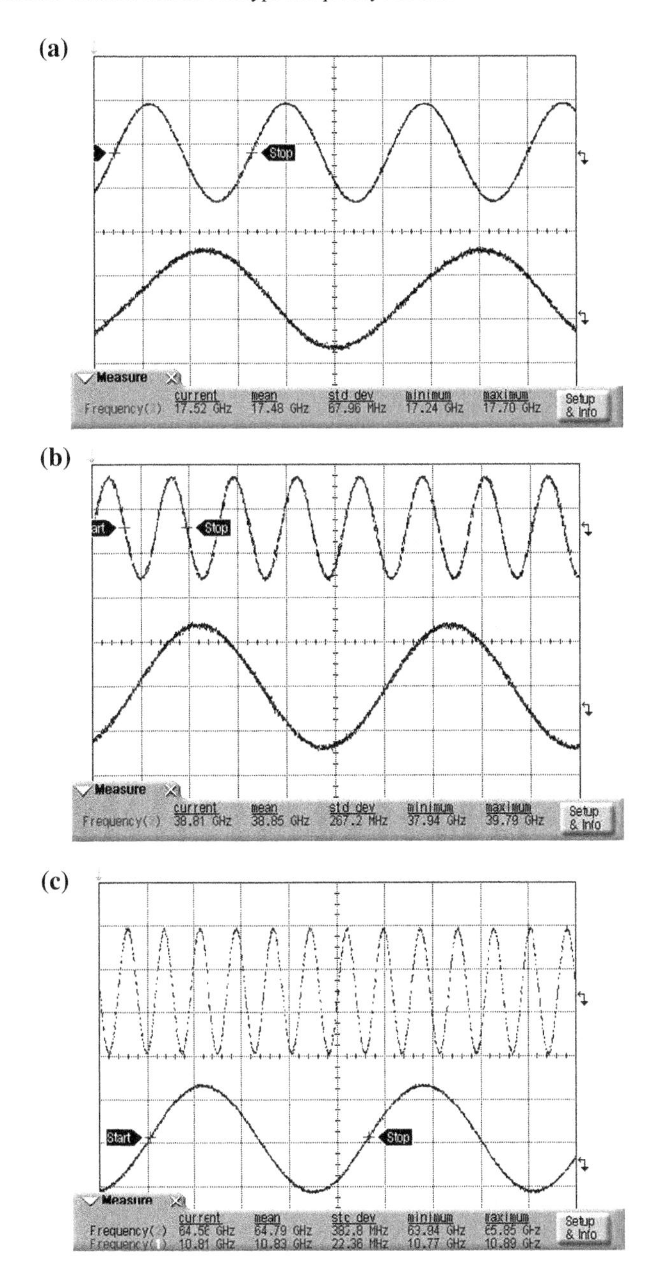

Fig. 3.6 Measured time domain waveforms when the ILFD is operating as **a** divide-by-two at 17 GHz, **b** divide-by-four at 39 GHz, and **c** divide-by-six at 65 GHz

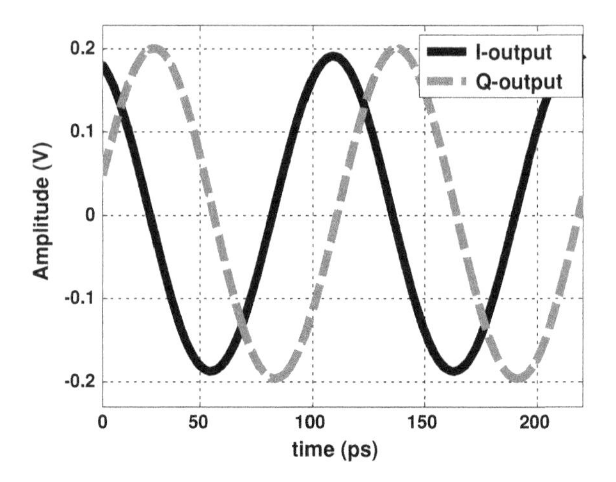

Fig. 3.7 Measured in-phase and quadrature phases at the output of ILFD operated in the divide-by-six mode, f_{in}=55 GHz

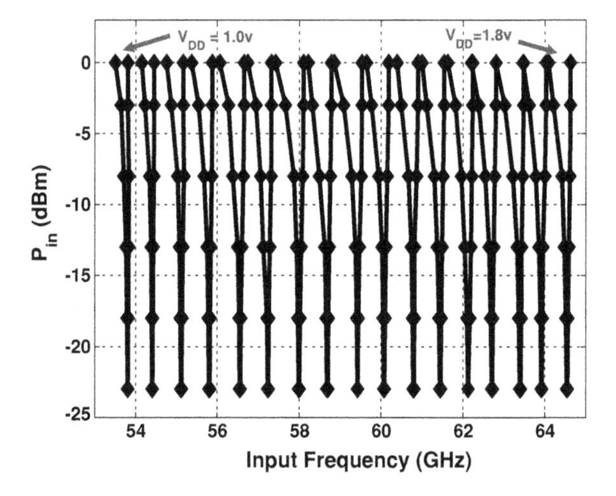

Fig. 3.8 Measured input frequency range for divide-by-six mode when external tuning is applied

The performance of this ILFD is compared with other published CMOS V-band frequency dividers, and is summarized in Table 3.1. Figure 3.11 shows the chip microphotograph.

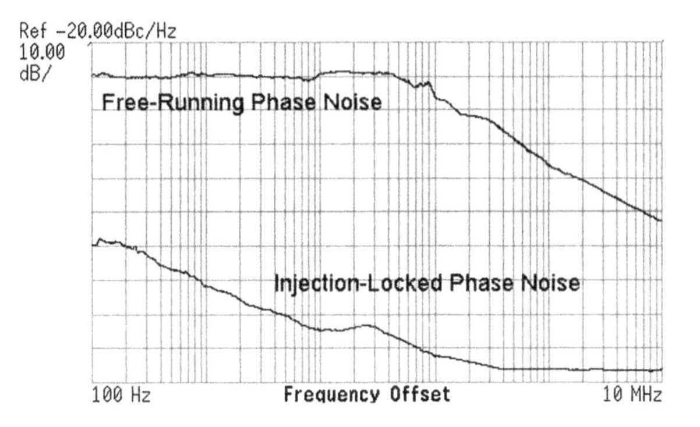

Fig. 3.9 Measured phase noise of the ILFD for free-running condition, and when ILFD is injection-locked to an input signal at 17.5 GHz

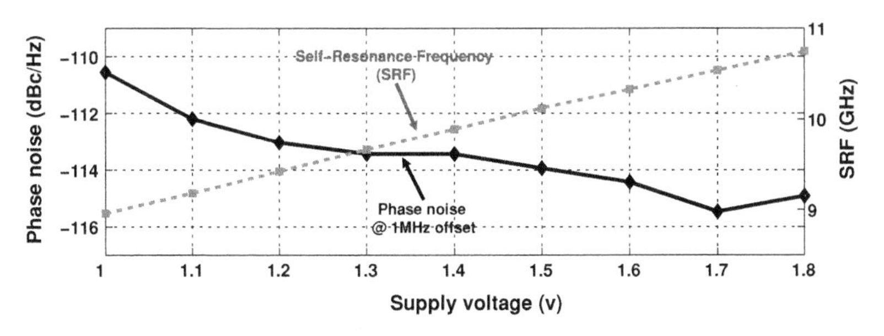

Fig. 3.10 Measured phase noise, and self-resonance frequency (SRF) for divide-by-six mode versus external tuning

Table 3.1 Performance comparison with recent 60–70 GHz band dividers.

Reference	[2]	[3]	This work
Input frequency (GHz)	70	70	65
Output phases	1	2	4 (Capable of 8)
Division ratio	4	2	6
Lock range (%)	1.3	9.4	1.5
Tuning range	63–72 GHz	No tuning	51–65 GHz
Input level (dBm)	0	−2	0
Phase noise (@ 1 MHz offset)	–	−114 dBc/Hz	−110 to −115 dBc/Hz
Technology (nm)	90	130	130
Supply current (mA)	5.5	5	12–24
Die size (mm^2)	0.014	0.120	0.026

3.5 Conclusion

A CMOS Millimeter-wave multi-phase divide-by-six ring-oscillator-based ILFD is presented. The divider also achieves division ratios of four and two when 44 or 22 GHz signals are applied, respectively. This ring-oscillator-based ILFD contains

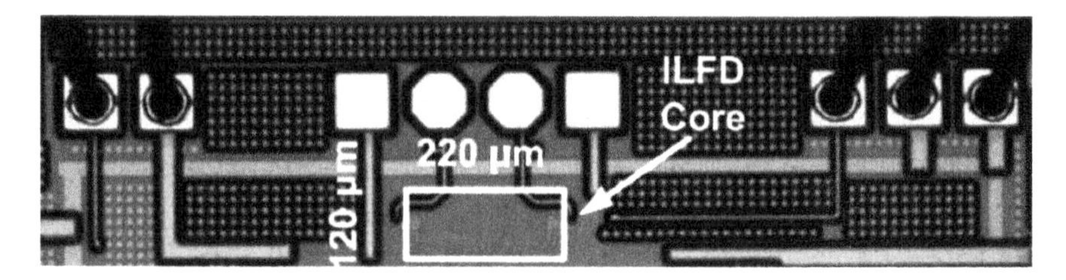

Fig. 3.11 Chip microphotograph

neither on-chip inductor or on-chip transformer, and the core area is 0.026 mm^2. This work demonstrates the possibility of designing compact, low-noise, multi-phase frequency dividers at frequencies close to (f_T) with CMOS technology.

References

1. Razavi B (1998), Architectures and circuits for RF CMOS receivers. In: Proceedings of the IEEE custom integrated circuits conference, pp 393–400, May 1998
2. Yamamoto K, Fujishima M (2006) 70 GHz CMOS harmonic injection-locked divider. In: IEEE ISSCC digest of technical papers, pp 2472–2481, Feb 2006
3. Razavi B (2007) Heterodyne phase locking: a technique for high-frequency division. In: IEEE ISSCC digest of technical papers, pp 428–429, Feb 2007
4. von Buren G, Kromer C, Ellinger F, Huber A, Schmatz M, Jackel H (2006) A combined dynamic and static frequency divider for a 40 GHz PLL in 80 nm CMOS. In: IEEE ISSCC digest of technical papers, pp 2462–2471, Feb 2006
5. Lee T (2004) The Design of CMOS radio-frequency integrated circuits. Cambridge University Press, Cambridge
6. Mirzaei A, Heidari M, Bagheri R, Abidi A (2008) Multi-phase injection widens lock range of ring-oscillator-based frequency dividers. IEEE J Solid-State Circuits 43(3):656–671
7. Betancourt-Zamora R, Verma S, Lee T (2001) 1-GHz and 2.8-GHz CMOS injection-locked ring oscillator prescalers. Symposium on VLSI Circuits Digital Technical Papers, pp 47–50, 2001
8. Farazian M, Gudem P, Larson L (2009) A CMOS multi-phase injection-locked frequency divider for V-band operation. IEEE Micro Wirel Compon Lett 14(1):447–450
9. Hossain M, Carusone A (2009) CMOS oscillators for clock distribution and injection-locked deskew. IEEE J Solid-State Circuits 44(8):2138–2153
10. Ghilioni A, Decanis U, Monaco E, Mazzanti A, Svelto F (2011) A 6.5 mW inductorless CMOS frequency divider-by-4 operating up to 70 GHz. In: IEEE ISSCC digest of technical papers, pp 282–284, Feb 2011
11. Pellerano S, Mukhopadhyay R, Ravi A, Laskar J, and Palaskas Y (2008) A 39.1–41.6 GHz $\Delta\Sigma$ fractional-N frequency synthesizer in 90 nm CMOS. In: IEEE ISSCC digest of technical papers, pp 484–630, Feb 2008
12. Scheir K, Vandersteen G, Rolain Y, Wambacq P (2009) A 57–66 GHz quadrature PLL in 45 nm digital CMOS. In: IEEE ISSCC digest of technical papers, pp 494–495, Feb 2009
13. Verma S, Rategh H, Lee T (2003) A unified model for injection-locked frequency dividers. IEEE J Solid-State Circuits 38(6):1015–1027

Chapter 4
Analysis of Injection-Locked Regenerative Frequency Dividers

4.1 Overview

Frequency dividers are one of the most important components of any frequency synthesizer. In some applications, fractional division ratios are required [1–3]. In addition, 50 % duty cycle quadrature output phases will allow the use of single-sideband (SSB) frequency conversion. It is challenging to implement a high-frequency fractional 50 % duty cycle and quadrature output frequency divider in a digital CMOS technology with small die area and low power consumption.

We begin this chapter with a general overview of fractional frequency divider architectures in Sect. 4.2. Injection-locked regenerative frequency dividers are introduced in Sect. 4.3. Their stability is analyzed in Sect. 4.4. Two-stage ring-oscillators are good candidates for quadrature output injection-locked regenerative frequency dividers. Their steady-state operation, stability, and injection-locked behavior (based on negative resistance delay cells) are studied in Sect. 4.5. Phase noise analysis of the injection-locked regenerative frequency dividers is performed in Sect. 4.6. Section 4.7 contains a design example of injection-locked regenerative dividers to implement a divider with fractional division ratio and quadrature output phases. Section 4.8 summarizes and concludes this chapter.

4.2 Fractional Frequency Dividers

Fractional frequency dividers are often implemented with multimodulus frequency dividers, and are used in fractional-N frequency synthesizers to implement a fractional division ratio [4]. In this technique, interpolating between two or more integer division ratios provides a division ratio that, on average, represents a fractional number [4]. Unfortunately, the output waveform achieved with this technique does not

M. Farazian et al., *Fast Hopping Frequency Generation in Digital CMOS*,
DOI: 10.1007/978-1-4614-0490-3_4, © Springer Science+Business Media New York 2013

exhibit a 50 % duty cycle when interpolating between different moduli of the frequency divider. Consequently, this output waveform can be used in applications such as fractional-N phase-locked loops, where the instantaneous period and the duty cycle of the signal are less important, and the loop filter averages the output of the phase detector (or charge pump). However, this waveform cannot be easily used for frequency translation in mixers, or in any application that is sensitive to the instantaneous period (or frequency) of the input signal. Hence, there is a need to develop frequency dividers that can *directly* generate fractional division ratios.

Regenerative frequency dividers are another technique to achieve fractional division ratios [5]. Figure 4.1a shows a block diagram of a regenerative frequency divider. The stability and operation of regenerative frequency dividers are studied in [5–8]. As shown in Fig. 4.1a, a mixer, or a nonlinear network in general, is used to create the mixing products of the input and output frequencies. The tuned network in the forward path passes the desired mixing product to the output. As shown in [5], this divider can achieve the desired fractional division ratio. However, regenerative frequency dividers have several limitations: (1) They are usually not able to provide quadrature output phases, (2) The sensitivity and output amplitude are degraded when higher order mixing products are needed to achieve larger than two, or fractional, division ratios [9], (3) They often require inductively tuned loads, which require a large die area, and (4) The locking range is limited by the Q of the tuned load.

The block diagram of a frequency divider based on the heterodyne phase locking technique [10] is shown in Fig. 4.1b. This technique can also be used to obtain a fractional division ratio. However, it requires tuned loads for filtering the sum component at the output of the mixer. In addition, the divider of Fig. 4.1b requires more than one mixer to implement a fractional division ratio. Increasing the number of mixers will increase the power consumption and die area. At the same time, this frequency divider cannot provide quadrature phases of the output, unless a quadrature VCO, or a combination of a VCO at twice the desired frequency and a divide-by-two is used in the forward path. These approaches require more area and power consumption. In addition, a quadrature VCO poses its own limitations on the I/Q amplitude and phase accuracy, as well as the achievable phase noise.

Another approach to implement a fractional division ratio is to use multiple phases of the input clock and interpolating between them using a sequential logic circuit [11]. However, the speed of this technique is limited to lower frequencies, since it relies on digital sequential circuits. Moreover, it can neither provide quadrature output phases nor achieve a 50 % duty cycle.

In this chapter, a different category of frequency dividers is analyzed, which can achieve both fractional division ratio and 50 % duty cycle quadrature output phases. This class of frequency dividers is discussed in Sect. 4.3, and its operation and stability conditions are analyzed in Sect. 4.4.

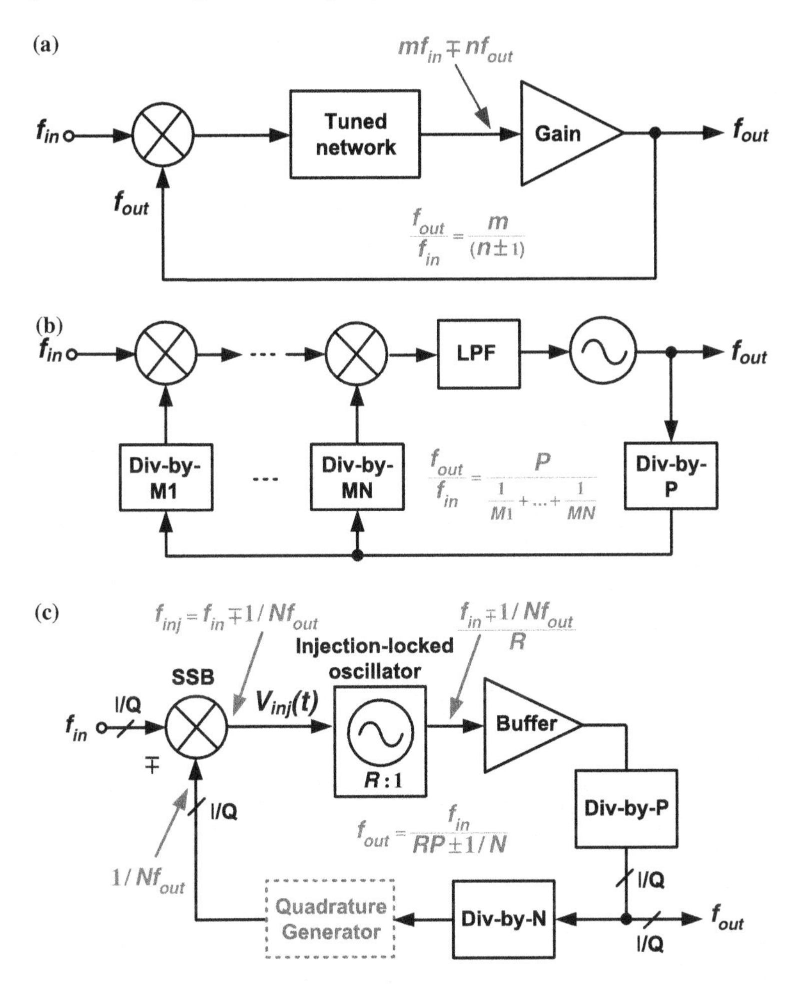

Fig. 4.1 **a** Block digram of a traditional regenerative frequency divider; **b** general block diagram of a frequency divider based on heterodyne phase locking technique [10]; **c** generalized injection-locked regenerative frequency divider

4.3 Injection-Locked Regenerative Frequency Dividers

A general block diagram of the injection-locked regenerative frequency divider is shown in Fig. 4.1c. This divider consists of an injection-locked oscillator in the forward path, which is followed by a frequency divider with division ratio P, and a frequency divider with division ratio N in the feedback path. The mixer in this divider functions as a frequency converter, and the signal at the output of the mixer has a component at $f_{\text{in}} \mp 1/Nf_{\text{out}}$. The mixer output, V_{inj}, can injection-lock the oscillator if its amplitude is sufficiently large and its frequency, f_{inj}, is within the locking range of the oscillator.

A SSB mixer can be used in this architecture to set the mixer output frequency to $(f_{\text{in}} - f_{\text{out}}/N)$ or $(f_{\text{in}} + f_{\text{out}}/N)$ if quadrature phases of both signals are available. This helps eliminate the inductor tuned load of the mixer, which eases the implementation of the frequency divider in a digital CMOS technology and expands the frequency divider input range.

If the oscillator is injection-locked to the Rth harmonic of its output frequency, i.e., $f_{\text{inj}} = Rf_{\text{out}}$, then the output frequency of the injection-locked regenerative frequency divider can be expressed by

$$f_{\text{out}} = \frac{f_{\text{in}}}{RP \pm \frac{1}{N}}. \tag{4.1}$$

The $\pm$ in (4.1) depends on whether the difference or sum of two frequencies is chosen at the SSB mixer output. Equation (4.1) shows the possibility of obtaining fractional division ratios using the injection-locked regenerative frequency divider of Fig. 4.1c. Similarly, if the oscillator is injection-locked to its Rth subharmonic, i.e., $f_{\text{inj}} = f_{\text{out}}/R$, then

$$f_{\text{out}} = \frac{f_{\text{in}}}{\frac{P}{R} \pm \frac{1}{N}}. \tag{4.2}$$

As can be seen from (4.1) and (4.2), injection-locked regenerative frequency dividers can generate almost any arbitrary division ratio. Moreover, as can be observed from Fig. 4.1c, it can simultaneously provide multiple division ratios. Additionally, the proper choice of the divide-by-P block, e.g. a divide-by-two, enables a 50 % duty cycle quadrature outputs. Appropriate distribution of the integer part of the division ratio in (4.1), i.e., RP (or the P/R ratio in (4.2)) may provide more options that can also lead to 50 % duty cycle quadrature outputs. A design example that achieves fractional division ratios, a 50 % duty cycle, and quadrature outputs is presented in Sect. 4.7.

Both (4.1) and (4.2) show the possibility of obtaining fractional division ratios. However, for the rest of this chapter our focus is on the case of super-harmonic injection-locking the oscillator to its Rth harmonic, i.e. $f_{\text{inj}} = Rf_{\text{out}}$, since this scenario has more applications in the frequency divider arena; hence we use (4.1) as the input–output frequency relationship of the frequency divider of Fig. 4.1.

Because of the similarities between the regenerative frequency divider of Fig. 4.1a and the injection-locked regenerative frequency divider, the divider of Fig. 4.1c is sometimes referred to as a *Modified Regenerative Divider* [3]. However, the divider of Fig. 4.1c is a separate category of frequency divider, since it relies on an injection-locked oscillator for its operation. To clarify this point, when there is no input signal to a traditional regenerative frequency divider, it does not generate any output. In fact, as discussed in [5], this is one of the stability criteria of regenerative frequency dividers. However, in the injection-locked regenerative frequency divider of Fig. 4.1c, the oscillator free-runs in the absence of an input.

4.4 Stability Analysis of Injection-Locked Regenerative Frequency Dividers

To analyze the stability of the frequency divider of Fig. 4.1c, we assume that the oscillator in the forward path is injection-locked to a frequency that is close to the Rth harmonic of its self-resonance frequency (SRF), and a SSB mixer is used to generate the difference (or sum) of f_{in} and $1/N f_{\text{out}}$. In this case, the relation between f_{out} and f_{in} is expressed by (4.1). In addition, we assume that the input signal is applied to the LO port of the SSB mixer. If we represent the input and output of the injection-locked regenerative frequency divider by $V_{\text{in}}(t) = A_i \cos(\omega_{\text{in}} t + \theta_i)$ and $V_{\text{out}}(t) = A_{\text{o}} \cos(\omega_{\text{out}} t)$, the differential mixer output, $V_{\text{inj}}(t)$, can be expressed as

$$V_{\text{inj}}(t) = G_M Z_M A_{\text{o}} \cos(\omega_{\text{out}} t/N) \times f[A_i \cos(\omega_{\text{in}} t + \theta_i)] \tag{4.3}$$

where G_M is the transconductance of the Gm stage of the SSB mixer, and Z_M is the load impedance of the SSB mixer. In an inductor-less design approach, Z_M is a parallel combination of the load resistor and the parasitic capacitances. Therefore, for simplicity, both Z_M and G_M can be considered constant within the locking range of the oscillator. Moreover, for simplicity, we assume that Z_M and G_M do not contribute any phase shift.

The function $f[\cdot]$ in (4.3) models the nonlinearity of the LO port of the mixer. If the input amplitude, A_i, is sufficiently large, $f[\cdot]$ can be approximated by a ± 1 square wave. Under this assumption, if we substitute the Fourier series expansion of $f[A_i \cos(\omega_{\text{in}} t + \theta_i)]$ into (4.3), the component of $V_{\text{inj}}(t)$ which is at a frequency close to the SRF of the oscillator is

$$V_{\text{inj}}(t) = \frac{2}{\pi} G_M Z_M A_{\text{o}} \cos\left(\frac{RP}{RP \pm \frac{1}{N}} \omega_{\text{in}} t + \theta_i \right). \tag{4.4}$$

Other mixing products at the output of the SSB mixer, which are caused by the nonlinearity of the LO port are sufficiently far from the SRF of the oscillator, hence cannot injection-lock the oscillator. Linearizing the Gm stage of the mixer will suppress the mixing products caused by the nonlinearity of its transconductance (G_M). As a result, only the term shown in (4.4) plays a role in injection-locking the oscillator and other terms are ignored.

Clearly, the minimum input sensitivity of the frequency divider occurs when the output frequency is f_{SRF}/P, which corresponds to $f_{\text{in}} = [RP \pm 1/N)/P] f_{\text{SRF}}$, where f_{SRF} is the SRF of the oscillator. If the amplitude of the signal at the mixer output (V_{inj}) is adequate to injection-lock the oscillator to $[P/(RP \pm 1/N)] f_{\text{in}}$, the oscillator and the frequency divider will operate in the stable region. If the amplitude is not adequate, the oscillator is pulled and will generate sidebands [12]. As a result, the stable region of operation of the injection-locked regenerative frequency divider is determined by the locking range of the oscillator.

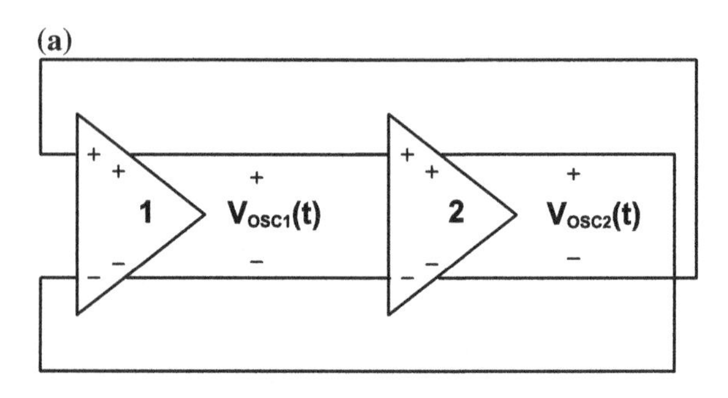

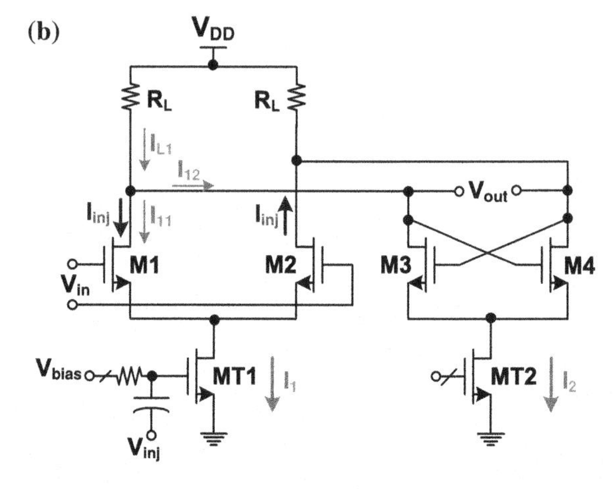

Fig. 4.2 Two-stage CMOS ring-oscillator. **a** Block diagram; **b** negative resistance delay cell

In order to super-harmonic injection-lock an oscillator to one of the oscillator's even harmonics, the injection signal must be applied to a common-mode node [13]. For instance, in the two-stage ring-oscillator of Fig. 4.2a, which can be implemented using the negative resistance delay cell of Fig. 4.2b, the injection current signal must be applied to the source terminals of transistors $M1$ and $M2$. In order to do that, a common choice is to apply $V_{\text{inj}}(t)$ to the gate of the tail current source of one of the delay cells in the ring-oscillator. If $V_{\text{inj}}(t)$ is applied to the gate of transistor $MT1$ of Fig. 4.2b, the component of injection current $(I_{\text{inj}}(t))$ at frequency $[P/(RP \pm 1/N)]f_{\text{in}}$ that reaches the oscillator output can be found using an approach similar to [13], i.e.,

$$I_{\text{inj}}(t) = \frac{8}{\pi^2(R^2-1)} g_{mT} G_M Z_M A_{\text{o}} \frac{\sin\theta_i}{\cos\chi} \times \cos\left(\frac{P}{RP \pm \frac{1}{N}}\omega_{\text{in}} t - \chi\right) \tag{4.5}$$

where

$$\chi = \tan^{-1}(R\cot\theta_i) \tag{4.6}$$

and g_{mT} is the transconductance of $MT1$ of Fig. 4.2b. The $I_{\text{inj}}(t)$ expressed in (4.5) injection-locks the oscillator to frequency $[P/(RP \pm 1/N)]f_{\text{in}}$. The amplitude of this injection current in terms of θ_i can be expressed as

$$\left|I_{\text{inj}}\right| = \frac{8}{\pi^2(R^2-1)} g_{mT} G_M Z_M A_o \sqrt{1+(R^2-1)\cos^2\theta_i}. \qquad (4.7)$$

These expressions are valid for differential oscillators where the injection signal is applied to the gate of the tail current source.

It is clear that the stable region of operation of an injection-locked regenerative frequency divider depends on the relationship between the locking range of the oscillator and the amplitude of the injection signal. This relationship is derived for LC oscillators in [12, 14], and for ring-oscillators, with more than three stages, in [15–17]. The locking range of a two-stage ring-oscillator is of interest, since it requires the fewest number of delay cells to generate quadrature output phases. Consequently, it can achieve smaller die area and lower power consumption. Therefore, the free-running and injection-locked behavior of this oscillator will be studied in Sect. 4.5, and its locking range for different injection-locking scenarios will be derived.

4.5 Injection-Locked Two-Stage Ring-Oscillator

In the previous section, we showed how injection at the gate of $MT1$ generates differential injection current at the output of oscillator, as shown in Fig. 4.3. Writing KCL at the drain of $M1$ of the first delay cell results in

$$\frac{V_{\text{OSC1\_}}(t) - V_{\text{DD}}}{R} + C\frac{\text{d}}{\text{d}t}V_{\text{OSC1\_}}(t) + I_{D1}(t) + I_{D3}(t) - I_{\text{inj\_}}(t) = 0. \qquad (4.8)$$

Similarly, writing KCL at the drain of transistor $M2$ of the first delay cell results in

$$\frac{V_{\text{OSC1}_+}(t) - V_{\text{DD}}}{R} + C\frac{\text{d}}{\text{d}t}V_{\text{OSC1}_+}(t) + I_{D2}(t) + I_{D4}(t) - I_{\text{inj}_+}(t) = 0. \qquad (4.9)$$

From (4.8) and (4.9) we obtain the following differential equation.

$$\frac{V_{\text{OSC1}}(t)}{R} + C\frac{\text{d}}{\text{d}t} = V_{\text{OSC1}}(t) = [I_{D1}(t) - I_{D2}(t)] + [I_{D3}(t) - I_{D4}(t)] + I_{\text{inj}}(t). \qquad (4.10)$$

In (4.10), $I_{\text{inj}}(t)$ is the differential injection current and is defined by $I_{\text{inj}_+}(t) - I_{\text{inj\_}}(t)$, and R and C are the equivalent output resistance and capacitance at the output of the delay cell. A similar differential equation is obtained for the second delay cell, without the injection current, as shown below.

$$\frac{V_{OSC2}(t)}{R} + C\frac{\mathrm{d}}{\mathrm{d}t}V_{OSC2}(t) = [I'_{D1}(t) - I'_{D2}(t)] + [I'_{D3}(t) - I'_{D4}(t)] \tag{4.11}$$

where $I'_{D1}(t)$, $I'_{D2}(t)$, $I'_{D3}(t)$, and $I'_{D4}(t)$ are the drain currents of the second delay cell.

In the steady-state, $V_{OSC1}(t)$ and $V_{OSC2}(t)$ can be expressed using Fourier series representation, i.e.

$$V_{OSC1}(t) = \sum_{k=-\infty}^{+\infty} V1_{(2k-1)}\mathrm{e}^{j\theta_{1,(2k-1)}(t)} \tag{4.12a}$$

$$V_{OSC2}(t) = \sum_{k=-\infty}^{+\infty} V2_{(2k-1)}\mathrm{e}^{j\theta_{2,(2k-1)}(t)} \tag{4.12b}$$

where

$$\theta_{1,(2k-1)}(t) = (2k-1)\omega_o t + \phi_{1,(2k-1)} \tag{4.13a}$$

$$\theta_{2,(2k-1)}(t) = (2k-1)\omega_o t + \phi_{2,(2k-1)}. \tag{4.13b}$$

In order to have a real solution for oscillator voltages, Vi_m and Vi_{-m} must be complex conjugates ($i = 1, 2$).

If the values of $V_{OSC1}(t)$ and $V_{OSC2}(t)$ are sufficiently large, the transistors of each delay cell are fully switched and the current waveforms of $I_{D1}(t) - I_{D2}(t)$ and $I_{D3}(t) - I_{D4}(t)$ are similar to a ± 1 square wave. These waveforms are in-phase with the fundamental component of their controlling voltages [15], i.e. $I_{D1}(t) - I_{D2}(t)$ in the first delay cell is in phase with the fundamental harmonic of $-V_{OSC2}(t)$ and $I_{D3}(t) - I_{D4}(t)$ is in phase with the fundamental harmonic of $V_{OSC1}(t)$, thus they could be represented using a Fourier series as follows.

$$[I_{D1}(t) - I_{D2}(t)] = \sum_{k=-\infty}^{+\infty} I1_{(2k-1)}e^{j[(2k-1)\theta_{2,1}(t)-\pi]} \tag{4.14a}$$

$$[I_{D3}(t) - I_{D4}(t)] = \sum_{k=-\infty}^{+\infty} I2_{(2k-1)}e^{j(2k-1)\theta_{1,1}(t)} \tag{4.14b}$$

The factor $-\pi$ in the argument of the instantaneous phase of $I_{D1}(t) - I_{D2}(t)$ in (4.14a) comes from its controlling voltage, which is the fundamental harmonic of $-V_{OSC2}(t)$. To represent a real current, $Ij_{(2k-1)}$ and $Ij_{-(2k-1)}$ must be complex conjugate ($j = 1, 2$). As shown in [18], under the stated assumptions, $I1_1$ and $I2_1$ are, respectively, $2I_1/\pi$ and $2I_2/\pi$. We also assume that the differential injection current can be represented by $I_{\mathrm{inj}}\cos(\theta_{\mathrm{inj}}(t))$ where $\theta_{\mathrm{inj}} = \omega_{\mathrm{inj}}t + \phi_{\mathrm{inj}}$. By substituting

(4.12a), (4.12b), (4.14a), and (4.14b) into (4.10), and considering stabilized amplitude of oscillation under steady-state, (4.15) is obtained.

$$\sum_{k=-\infty}^{+\infty} \frac{V1_{(2k-1)}}{R} e^{j\theta_{1,(2k-1)}(t)} + jC\left(\sum_{k=-\infty}^{+\infty} V1_{(2k-1)} \frac{\mathrm{d}}{\mathrm{d}t}[\theta_{1,(2k-1)}(t)] e^{j\theta_{1,(2k-1)}(t)}\right)$$
$$= -\sum_{k=-\infty}^{+\infty} I1_{(2k-1)} e^{j(2k-1)\theta_{2,1}(t)} + \sum_{k=-\infty}^{+\infty} I2_{(2k-1)} e^{j(2k-1)\theta_{1,1}(t)} + I_{\mathrm{inj}} \cos\theta_{\mathrm{inj}} \tag{4.15}$$

When the oscillator is injection-locked, $\omega_o = \omega_{\mathrm{inj}}$. Equating the coefficients of similar exponents in (4.15) results in (4.16) where $\delta_{(2k-1),\pm 1}$ is the Kronecker delta and equals unity for the fundamental harmonic of output voltages ($k = 0, 1$) and is zero for other harmonics.

$$\frac{V1_{(2k-1)}}{R} e^{j\theta_{1,(2k-1)}(t)} + jCV1_{(2k-1)} \frac{\mathrm{d}}{\mathrm{d}t}[\theta_{1,(2k-1)}(t)] e^{j\theta_{1,(2k-1)}(t)}$$
$$= -I1_{(2k-1)} e^{j(2k-1)\theta_{2,1}(t)} + I2_{(2k-1)} e^{j(2k-1)\theta_{1,1}(t)} + \frac{1}{2} I_{\mathrm{inj}} e^{j\theta_{\mathrm{inj}}(t)} \delta_{(2k-1),\pm 1} \tag{4.16}$$

where $k \in \mathbb{Z}$.

The harmonics of the currents of (4.14a) and (4.14a) have a roll-off of approximately $1/|2k-1|$. Moreover, these harmonics go through the low-pass filter of the output load of each delay cell. As a result, the fundamental harmonic is dominant.

Substituting $k = 1$ into (4.16) results in a differential equation for the fundamental harmonic of $V_{\mathrm{OSC1}}(t)$,

$$\frac{V1_1}{R} e^{j\theta_{1,1}(t)} + jCV1_1 \frac{\mathrm{d}\theta_{1,1}(t)}{\mathrm{d}t} e^{j\theta_{1,1}(t)}$$
$$= -I1_1 e^{j\theta_{2,1}(t)} + I2_1 e^{j\theta_{1,1}(t)} + \frac{1}{2} I_{\mathrm{inj}} e^{j\theta_{\mathrm{inj}}(t)} \tag{4.17}$$

After substituting the values of $I1_1$ and $I2_1$ into (4.17), it can be rewritten as

$$\frac{V1_1}{R} + jCV1_1 \frac{\mathrm{d}\theta_{1,1}(t)}{\mathrm{d}t} = -\frac{2I_1}{\pi} e^{j[\theta_{2,1}(t)-\theta_{1,1}(t)]}$$
$$+ \frac{2I_2}{\pi} + \frac{1}{2} I_{\mathrm{inj}} e^{j[\theta_{\mathrm{inj}}(t)-\theta_{1,1}(t)]} \tag{4.18}$$

A similar equation can be obtained for the second delay cell, as follows.

$$\frac{V2_1}{R} + jCV2_1 \frac{\mathrm{d}\theta_{2,1}(t)}{\mathrm{d}t} = \frac{2I_1}{\pi} e^{j[\theta_{1,1}(t)-\theta_{2,1}(t)]} + \frac{2I_2}{\pi} \tag{4.19}$$

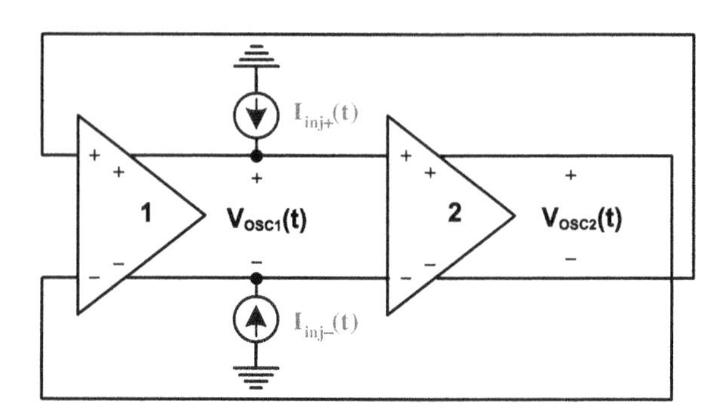

Fig. 4.3 Schematic of the two-stage CMOS ring-oscillator when an external signal is injected at the output of the first delay cell

To further simplify our analysis, we define $\Delta\theta$ and ψ as follows

$$\Delta\theta \triangleq \theta_{1,1}(t) - \theta_{2,1}(t) \quad (4.20a)$$

$$\psi \triangleq \theta_{\text{inj}}(t) - \theta_{1,1}(t). \quad (4.20b)$$

By separating the real and imaginary parts of (4.18) and (4.19) we obtain

$$\frac{d\theta_{1,1}(t)}{dt} = \frac{1}{RC}\frac{I_1 \sin\Delta\theta + \frac{\pi}{4} I_{\text{inj}} \sin\psi}{-I_1\cos\Delta\theta + I_2 + \frac{\pi}{4} I_{\text{inj}}\cos\psi} \quad (4.21a)$$

$$\frac{d\theta_{2,1}(t)}{dt} = \frac{1}{RC}\frac{I_1 \sin\Delta\theta}{I_1\cos\Delta\theta + I_2}. \quad (4.21b)$$

These nonlinear differential equations are very similar to those from [15] for ring-oscillators with more than three stages. Equations (4.18) and (4.19) are also used to calculate $V1_1$ and $V2_1$. If we represent the amplitudes of the fundamental harmonics of the first and second stage by V_{a1} and V_{a2}, $V_{a1} = 2\text{Re}[V1_1]$ and $V_{a2} = 2\text{Re}[V2_1]$, thus

$$V_{a1} = R\left(-\frac{4I_1}{\pi}\cos\Delta\theta + \frac{4I_2}{\pi} + I_{\text{inj}}\cos\psi\right) \quad (4.22a)$$

$$V_{a2} = R\left(\frac{4I_1}{\pi}\cos\Delta\theta + \frac{4I_2}{\pi}\right). \quad (4.22b)$$

We use (4.21a) and (4.21b) to analyze the free-running and injection-locking behavior of the two-stage ring-oscillator of Fig. 4.3.

4.5.1 Free-Running Oscillation

In steady-state and in the absence of an external signal, i.e., $I_{\text{inj}} = 0$, the ring-oscillator oscillates at its SRF. In this case,

$$\frac{d\theta_{1,1}(t)}{dt} = \frac{d\theta_{2,1}(t)}{dt} = \omega_{SRF}. \tag{4.23}$$

After substituting $I_{inj} = 0$ into (4.21a) and combining it with (4.21b) and (4.23), and from Appendix A, it is concluded that in steady-state

$$\Delta\theta = +\frac{\pi}{2}. \tag{4.24}$$

The SRF is obtained by substituting $\Delta\theta = +\pi/2$ into (4.21a) or (4.21b), and then

$$\omega_{SRF} = \frac{1}{RC} \cdot \frac{I_1}{I_2} \tag{4.25}$$

and hence the steady-state solution for $\theta_{1,1}(t)$ and $\theta_{2,1}(t)$ can be written as shown below.

$$\theta_{1,1}(t) = \omega_{SRF}t \tag{4.26a}$$

$$\theta_{2,1}(t) = \omega_{SRF}t - \frac{\pi}{2} \tag{4.26b}$$

From (4.22a), (4.22b), and (4.24) the steady-state amplitudes of the fundamental harmonics of the output voltages are

$$V_{a1} = V_{a2} = \frac{4I_2R}{\pi}. \tag{4.27}$$

4.5.2 Two-Stage Ring-Oscillator Under Single Node Injection

Assuming that an external signal is injected at the output of the first delay cell of the two-stage ring-oscillator, as shown in Fig. 4.3, and has injection-locked the oscillator to its frequency (ω_{inj}). In this case,

$$\frac{d\theta_{1,1}(t)}{dt} = \frac{d\theta_{2,1}(t)}{dt} = \omega_{inj}. \tag{4.28}$$

We can use (4.28), (4.21a), and (4.21b) to find the steady-state solution for $\theta_{1,1}$ and $\theta_{2,1}$. Also, from (4.21a), (4.21b), and (4.28) the oscillation frequency under injection-locking can be expressed as

$$\omega_o|_{inj} = \omega_{inj} = \frac{1}{RC} \frac{I_1 \sin\Delta\theta}{I_1 \cos\Delta\theta + I_2}. \tag{4.29}$$

It is important to note that the oscillation frequency for this scenario is a function of $\Delta\theta$. Substituting the ω_{SRF} from (4.25) into (4.29) results in

$$\omega_{inj} = \omega_{SRF} \frac{\sin \Delta\theta}{1 + RC\omega_{SRF} \cos \Delta\theta}. \tag{4.30}$$

The corresponding phase shift ($\Delta\theta$) for any given injection frequency (ω_{inj}) can be found from (4.30). Equation (4.30) can be re-written as

$$1 + \alpha \cos \Delta\theta - \beta \sin \Delta\theta = 0 \tag{4.31}$$

where α and β are

$$\alpha = RC\omega_{SRF} \tag{4.32a}$$

$$\beta = \frac{\omega_{SRF}}{\omega_{inj}}. \tag{4.32b}$$

The solution for (4.31) can be expressed as

$$\Delta\theta = \sin^{-1}\left(\frac{1}{\sqrt{\alpha^2 + \beta^2}}\right) - \tan^{-1}\left(\frac{\alpha}{\beta}\right). \tag{4.33}$$

Equation (4.33) determines the phase difference between the fundamental component of the output voltages (V_{OSC1} and V_{OSC2}). As can be seen from (4.33), injection-locking a two-stage ring-oscillator to any frequency other than its SRF (using this scheme of injection-locking) results in non-quadrature fundamental harmonics of the outputs. As an example, a two-stage ring-oscillator, based on the delay cell of Fig. 4.2b is designed in a 0.13 μm CMOS technology using a 1.2 V supply, and is used to verify this conclusion through simulation. In this oscillator, $R_L = 500\,\Omega$, $I_1 = 1.25$ mA, $I_2 = 650$ μA, and the SRF is approximately 4 GHz. The calculated output phase difference for a prototype two-stage ring-oscillator is plotted in Fig. 4.4 and is compared to circuit simulation, and the calculation error is less than two degrees over the entire locking range.

We now calculate the required minimum amplitude and phase of the external signal to injection-lock the oscillator to ω_{inj}. From (4.21a), (4.21b), and (4.28) it can be concluded that

$$\frac{I_1 \sin \Delta\theta + \frac{\pi}{4} I_{inj} \sin \psi}{-I_1 \cos \Delta\theta + I_2 + \frac{\pi}{4} I_{inj} \cos \psi} = \frac{I_1 \sin \Delta\theta}{I_1 \cos \Delta\theta + I_2}. \tag{4.34}$$

Equation (4.34) can be simplified to

$$I_1^2 \sin 2\Delta\theta + \frac{\pi}{4} I_{inj} \left[I_1 \sin(\psi - \Delta\theta) + I_2 \sin \psi\right] = 0. \tag{4.35}$$

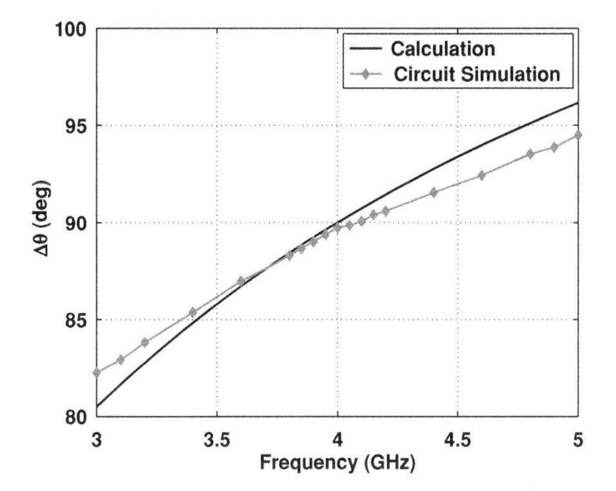

Fig. 4.4 Circuit simulated and calculated phase difference, using (4.33), at the output of a two-stage ring-oscillator, when an external signal is injected to the output of the first delay cell, $f_{\mathrm{SRF}} = 4\,\mathrm{GHz}$

Using (4.35) one can find ψ, as a function of $\Delta\theta$ and injection current, as follows

$$\psi = \xi - \sin\left(\frac{4}{\pi}\frac{I_1^2 \sin 2\Delta\theta}{I_{\mathrm{inj}}\sqrt{I_1^2 + I_2^2 + 2I_1 I_2 \cos\Delta\theta}}\right) \tag{4.36}$$

where

$$\xi = \tan^{-1}\left(\frac{I_1 \sin\Delta\theta}{I_1 \cos\Delta\theta + I_2}\right). \tag{4.37}$$

The smallest required amplitude of injection current to injection-lock the oscillator at ω_{inj} is obtained from (4.36).

$$I_{\mathrm{inj}} \geq \frac{4}{\pi}\frac{I_1^2\,|\sin 2\Delta\theta|}{\sqrt{I_1^2 + I_2^2 + 2I_1 I_2 \cos\Delta\theta}} \tag{4.38}$$

Therefore, by substituting $\Delta\theta$ from (4.33) into (4.38), one can find a lower bound for I_{inj} to injection-lock the oscillator to ω_{inj}. Lastly, the solution for ψ is found by substituting the lower bound for I_{inj} into (4.36).

Repeating this procedure for different values of ω_{inj} results in the input sensitivity curve (I_{inj} vs. ω_{inj}) of the two-stage ring-oscillator.

Using this procedure, the calculated locking range of the two-stage ring-oscillator is obtained and plotted in Fig. 4.5 and is compared with circuit simulations. It can be seen from Fig. 4.5 that the error in predicting the minimum injection current at the boundaries of the locking range is less than 6 %.

It can be concluded from (4.22a) and (4.22b) that in the presence of any external signal, even if it is at the same frequency as SRF, the amplitudes of the oscillation

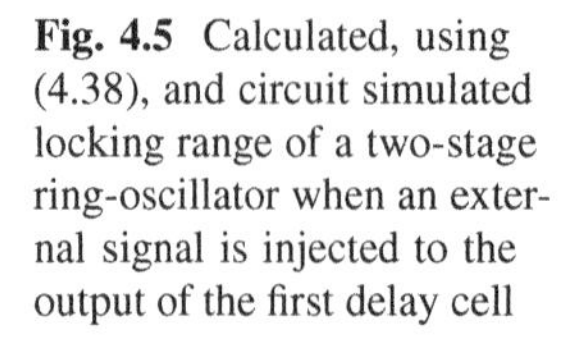

Fig. 4.5 Calculated, using (4.38), and circuit simulated locking range of a two-stage ring-oscillator when an external signal is injected to the output of the first delay cell

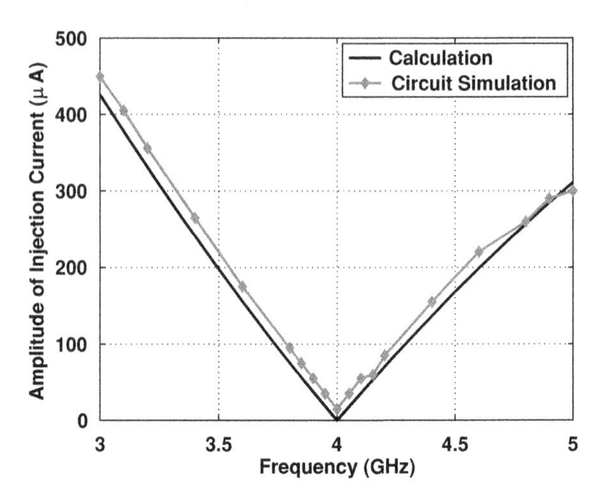

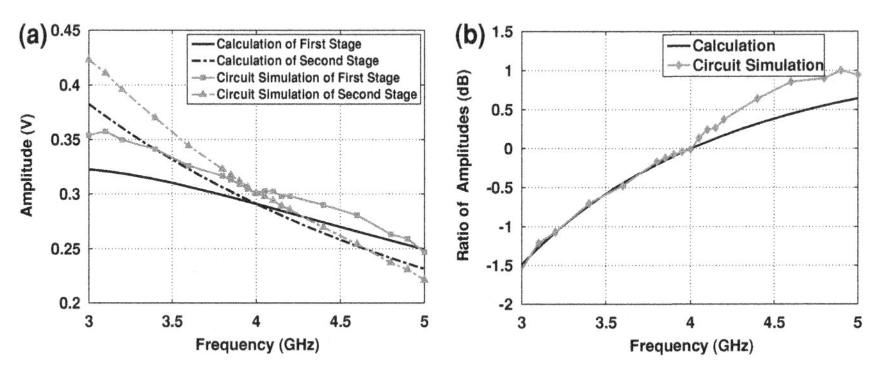

Fig. 4.6 Amplitudes of oscillation versus injection frequency **a** calculated, using (4.22a) and (4.22b), and circuit simulated amplitudes of output voltages, and **b** calculated and circuit simulated amplitude difference when external signal is injected only at the output of the first delay cell, and the minimum required injection current is applied, $f_{\mathrm{SRF}} = 4\,\mathrm{GHz}$.

voltages are not equal. Figure 4.6a shows the calculated and simulated amplitudes of the output voltages as a function of injection frequency, assuming the minimum required injection current, from (4.38), is injected at the output of the first delay cell. As can be observed in Fig. 4.6a, the output voltages have equal amplitude at the SRF. The calculated and simulated amplitude difference versus injection frequency are shown in Fig. 4.6b.

Figure 4.7a shows a graphical representation of the steady-state solution for the oscillation phases of the oscillator of Fig. 4.3 when it free-runs. In this representation, I_{11} and I_{12} are the corresponding phasors for the I_1 and I_2 current sources of the first delay cell, shown in Fig. 4.2b, and I_{21} and I_{22} are the corresponding current phasors of the second delay cell. The resultant currents of the delay cells are denoted by I_{L1} and I_{L2}. As can be seen in Fig. 4.7a, when the two-stage ring-oscillator free-runs,

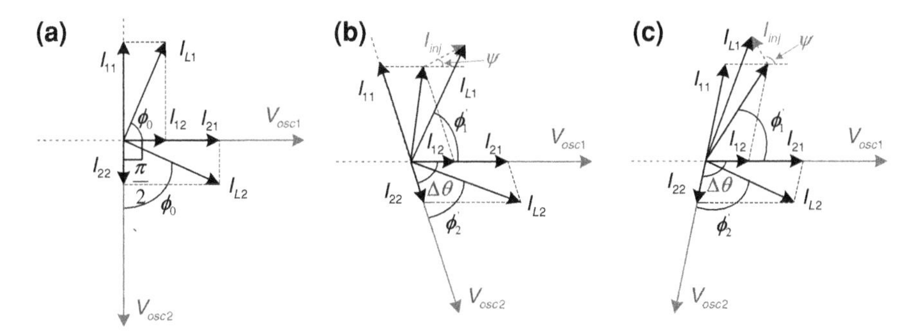

Fig. 4.7 Graphical representation of steady-state solution for the voltage and current phasors of the two-stage ring-oscillator. **a** Free-running; **b** injection-locked to a frequency lower than its self-resonance frequency (SRF), and **c** injection-locked to a frequency greater than its SRF. External signal is injected at the output of the first delay cell

or locks to its SRF, the two oscillation phases are orthogonal. In this case, the angle between the phasors of the resultant current and the corresponding voltage of each stage is $\phi_0 = \tan^{-1}(I_1/I_2)$.

Figure 4.7b shows the same currents and voltages when the oscillator of Fig. 4.3 is injection-locked to a frequency lower than its SRF. In this case, the phase difference between oscillation phases, $\Delta\theta$, is less than $\pi/2$. It can be shown that—in this case—the angle between the phasors of the resultant current and the voltage of each delay cell (ϕ_1' in the first delay cell and ϕ_2' in the second one) is less than ϕ_0. Similarly, Fig. 4.7c depicts the voltage and current phasors when the oscillator is injection-locked to a frequency greater than its SRF. In this case, $\Delta\theta$ is greater than $\pi/2$. Similarly, it can be shown that in this case ϕ_1' and ϕ_2' are greater than ϕ_0.

The non-quadrature output phases obtained for this scheme of injection-locking when the oscillator is injection-locked to frequencies other than its SRF, makes this scheme of injection-locking less attractive for the applications with demanding quadrature accuracy. In Sect. 4.5.3, this oscillator is analyzed when the external signal is injected at the output of both the delay cells.

4.5.3 Two-Stage Ring-Oscillator Multi-Node Injection

In this section, we investigate the two-stage ring-oscillator when the external signal is injected to both of its delay cells, as shown in Fig. 4.8. There are several reasons to inject the external signal at multiple nodes instead of a single node:

1. It provides balanced loading for the previous stage in differential circuits.
2. It may increase the locking range of the oscillator under injection, as shown in [15, 16].

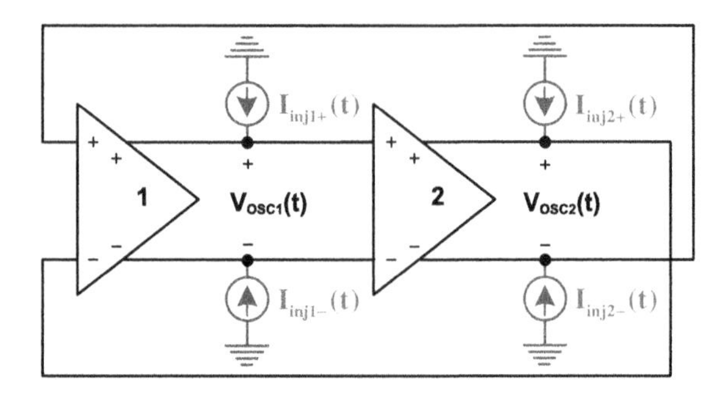

Fig. 4.8 Two-stage ring-oscillator when external signal is injected at the output of both delay cells

3. It may help to maintain quadrature phases ($\Delta\theta = \pi/2$) and equal amplitudes for the fundamental harmonics of output voltages in the entire locking range.

In this section, we propose a technique to maintain $\Delta\theta = \pi/2$ for the entire locking range of the oscillator.

Using similar assumptions and procedures that were used to derive (4.18) and (4.19), the following set of differential equations are derived for the amplitudes and phases of the oscillator when external signals are injected to the outputs of both delay cells (Fig. 4.8).

$$\frac{V1_1}{R} + jCV1_1\frac{d\theta_{1,1}(t)}{dt} = -\frac{2I_1}{\pi}e^{j[\theta_{2,1}(t)-\theta_{1,1}(t)]} + \frac{2I_2}{\pi} + \frac{1}{2}I_{\text{inj1}}e^{j[\theta_{\text{inj1}}(t)-\theta_{1,1}(t)]} \tag{4.39a}$$

$$\frac{V2_1}{R} + jCV2_1\frac{d\theta_{2,1}(t)}{dt} = \frac{2I_1}{\pi}e^{j[\theta_{1,1}(t)-\theta_{2,1}(t)]} + \frac{2I_2}{\pi} + \frac{1}{2}I_{\text{inj2}}e^{j[\theta_{\text{inj2}}(t)-\theta_{2,1}(t)]} \tag{4.39b}$$

Separating the real and imaginary parts of (4.39a) and (4.39b) will result in differential equations that relate the amplitudes and phases of the fundamental harmonic of the output voltages to the amplitudes and phases of the injection signals. One can use (4.20a) and the following definitions to simplify the results;

$$\psi_1 \triangleq \theta_{\text{inj1}}(t) - \theta_{1,1}(t) \tag{4.40a}$$

$$\psi_2 \triangleq \theta_{\text{inj2}}(t) - \theta_{2,1}(t) \tag{4.40b}$$

results in

$$\frac{d\theta_{1,1}(t)}{dt} = \frac{1}{RC}\frac{I_1\sin\Delta\theta + \frac{\pi}{4}I_{\text{inj1}}\sin\psi_1}{-I_1\cos\Delta\theta + I_2 + \frac{\pi}{4}I_{\text{inj1}}\cos\psi_1} \tag{4.41a}$$

$$\frac{d\theta_{2,1}(t)}{dt} = \frac{1}{RC}\frac{I_1 \sin \Delta\theta + \frac{\pi}{4} I_{inj2} \sin \psi_2}{I_1 \cos \Delta\theta + I_2 + \frac{\pi}{4} I_{inj2} \cos \psi_2}. \tag{4.41b}$$

In addition, the steady-state amplitudes of the fundamental harmonics of the output voltages are obtained from (4.39a) and (4.39b).

$$V_{a1} = R\left(-\frac{4I_1}{\pi}\cos \Delta\theta + \frac{4I_2}{\pi} + I_{inj1}\cos \psi_1\right) \tag{4.42a}$$

$$V_{a2} = R\left(\frac{4I_1}{\pi}\cos \Delta\theta + \frac{4I_2}{\pi} + I_{inj2}\cos \psi_2\right) \tag{4.42b}$$

To complete the analysis, we also assume that the external signals that are used to injection-lock the oscillator have equal amplitudes, i.e. $I_{inj1} = I_{inj2} = I_{inj}$. From (4.28), (4.41a), and (4.41b), it is concluded that in steady-state the following relation between ψ_1, ψ_2, $\Delta\theta$, and I_{inj} holds.

$$\frac{I_1 \sin \Delta\theta + \frac{\pi}{4} I_{inj} \sin \psi_1}{-I_1 \cos \Delta\theta + I_2 + \frac{\pi}{4} I_{inj} \cos \psi_1} = \frac{I_1 \sin \Delta\theta + \frac{\pi}{4} I_{inj} \sin \psi_2}{I_1 \cos \Delta\theta + I_2 + \frac{\pi}{4} I_{inj} \cos \psi_2} \tag{4.43}$$

We need to find a solution for (4.43) that satisfies $\Delta\theta = \pi/2$ for every value of I_{inj}, ψ_1, and ψ_2. By substituting $\Delta\theta = \pi/2$ into (4.43) we obtain the following equation.

$$\sin\left(\frac{\psi_1 - \psi_2}{2}\right)\left[2I_1 \sin\left(\frac{\psi_1 + \psi_2}{2}\right) + 2I_2 \cos\left(\frac{\psi_1 + \psi_2}{2}\right) + \frac{\pi}{2} I_{inj} \cos\left(\frac{\psi_1 - \psi_2}{2}\right)\right] = 0 \tag{4.44}$$

By inspection we can see that $\psi_1 = \psi_2$ satisfies (4.44) for all the values of I_{inj}. If we define $\Delta\theta_{inj}$ as $\theta_{inj1} - \theta_{inj2}$, it can be shown that to have quadrature output phases in the oscillator of Fig. 4.8, the following condition needs to be true.

$$\Delta\theta_{inj} = \pi/2 \tag{4.45}$$

To provide quadrature inputs for this scheme of injection-locking, one can use a polyphase filter, or supply these phases using another ring-oscillator. However, when the ring-oscillator of Fig. 4.8 is used as a divide-by-two frequency divider, the external signal is applied to the gate of tail current source $MT1$ of the delay cell shown in Fig. 4.2b. Since this signal is at a frequency twice the ω_{SRF}, the external signals applied to the gate terminals of the tail current sources in delay cells need to be 180 degrees out of phase to satisfy (4.45). This simplifies the problem of providing the oscillator under locking with the appropriate phases of external signal. Moreover, in this case, the oscillator/ frequency divider can provide balanced loading for its preceding differential stage. It is particularly of practical interest in injection-locked regenerative frequency dividers where the oscillator is driven by a differential SSB mixer.

By substituting the solution to (4.44), i.e. $\psi_1 = \psi_2 = \psi$, into (4.43) one can find an expression for the oscillation frequency under injection-locking in terms of the circuit parameters and the amplitudes and phases of the external signals.

$$\omega_{\text{inj}} = \omega_{\text{SRF}} \left(\frac{1 + \gamma \sin \psi}{1 + \delta \cos \psi} \right) \tag{4.46}$$

where

$$\gamma = \frac{\pi}{4} \frac{I_{\text{inj}}}{I_1} \tag{4.47a}$$

$$\delta = \frac{\pi}{4} \frac{I_{\text{inj}}}{I_2}. \tag{4.47b}$$

A comparison with the case where the external signal is only injected to the first delay cell reveals that the oscillation frequency of the multi-node injection case is modulated by the angle between the injection current and the oscillation voltage (ψ), as shown in (4.46), while $\Delta\theta$ modulates the oscillation frequency in the single-injection case, as shown in (4.30).

If ω_{inj} is in the locking range of the oscillator, ψ can be determined by

$$\psi = \sin^{-1} \left(\frac{\omega_{\text{inj}}/\omega_{\text{SRF}} - 1}{\gamma \sqrt{1 + \left(RC\omega_{\text{inj}}\right)^2}} \right) - \tan^{-1}(RC\omega_{\text{inj}}) \tag{4.48}$$

The minimum required injection current to injection-lock the two-stage ring -oscillator to ω_{inj} is obtained from (4.48) and (4.25).

$$I_{\text{inj}} \geq \frac{4}{\pi} \frac{I_1}{\sqrt{1 + \left(\frac{I_1}{I_2} \frac{\omega_{\text{inj}}}{\omega_{\text{SRF}}}\right)^2}} \left| \frac{\omega_{\text{inj}}}{\omega_{\text{SRF}}} - 1 \right| \tag{4.49}$$

Using (4.49) one can obtain the locking-range of the two-stage ring-oscillator when injection current is injected to both the delay cells. The calculated and simulated locking-range of the two-stage ring-oscillator are plotted in Fig. 4.9.

As can be seen from this figure, the calculation and simulations match within the locking range. They deviate at the high end of the locking range and the maximum error at 6.2 GHz is approximately 20 %, due to additional parasitic components that are not included in the model of the ILFD.

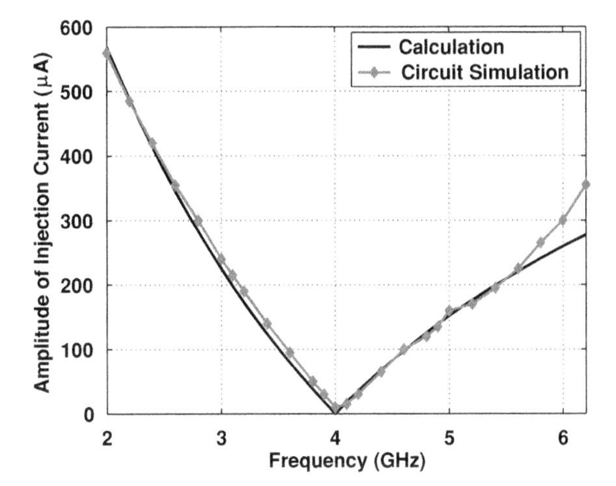

Fig. 4.9 Calculated, using (4.49), and circuit simulated locking range of a two-stage ring-oscillator when external signals are injected at the output of both delay cells

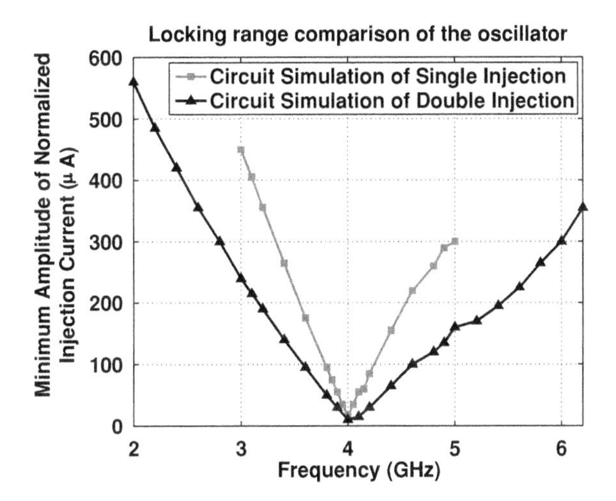

Fig. 4.10 Circuit simulated locking range of a two-stage ring-oscillator, $f_{SRF} = 4\,\text{GHz}$ for different injection-locking schemes

This simulated locking range is compared with the simulated locking range from Sect. 4.5.2 where the external signal was only applied to the first delay cell. The result vs. the normalized injection current amplitude is plotted in Fig. 4.10. It is observed from this plot that applying the external signal to both the delay cells, with appropriate phase sequence and equal amplitudes, leads to a wider locking range, and hence improves the sensitivity of the injection-locked oscillator. This result is in agreement with what was obtained in [15, 16] for ring-oscillators with more than three stages.

An output phase difference ($\Delta\theta$) of $\pi/2$ between fundamental harmonics of the output voltages is expected for this scheme of injection-locking, and the circuit simulations agree with that.

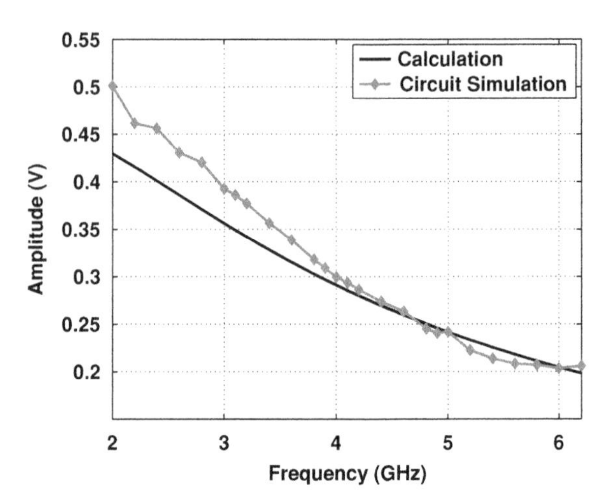

Fig. 4.11 Calculated, using (4.50), and circuit simulated amplitudes of output voltages versus injection frequency

Note that under these assumptions, the amplitudes of the output voltages remain equal for all the values of injection frequency within the locking range of the oscillator and are expressed by

$$V_{a1} = V_{a2} = R\left(\frac{4I_2}{\pi} + I_{\text{inj}} \cos \psi\right). \tag{4.50}$$

The calculated and simulated amplitudes of the fundamental component of the output voltages are plotted in Fig. 4.11. It is expected from (4.50) that for this scheme of injection-locking, the amplitudes of the output voltages remain equal within the locking range of the two-stage ring-oscillator, and this is confirmed in circuit simulations.

Figure 4.12 shows a graphical representation of the steady-state solution for the phasors of voltages and currents of the oscillator of Fig. 4.8 when external signals are injected to both delay cells. It can be seen that under the constraints derived for external signals ($I_{\text{inj}1} = I_{\text{inj}2}$ and $\Delta\theta_{\text{inj}} = \pi/2$) the oscillation voltages remain in quadrature with equal amplitudes. Similar to Fig. 4.7a, I_{11} and I_{12} are the corresponding phasors for I_1 and I_2 current sources (Fig. 4.2a) of the first delay cell, and I_{21} and I_{22} are the corresponding current phasors to the second delay cell, and I_{L1} and I_{L2} are the resultant currents of the first and second delay cells respectively.

Circuit simulations across the locking range show that a 20 % input amplitude mismatch leads to quadrature amplitude and phase mismatches of 0.4 dB and 2° at the output of the ring-oscillator of Fig. 4.8. In this case, the amplitudes of the injection currents injected to the outputs of the delay cells of Fig. 4.8 are 605 and 495 μA. On the other hand, an input phase mismatch of ±5° leads to output amplitude and phase mismatches of 0.4 dB and 0.5° respectively. The injection-locked oscillator draws 4 mA from a 1.2 V supply across the locking range. It can be shown that, if the input amplitude mismatch is negligible, an input phase mismatch of ε causes an output

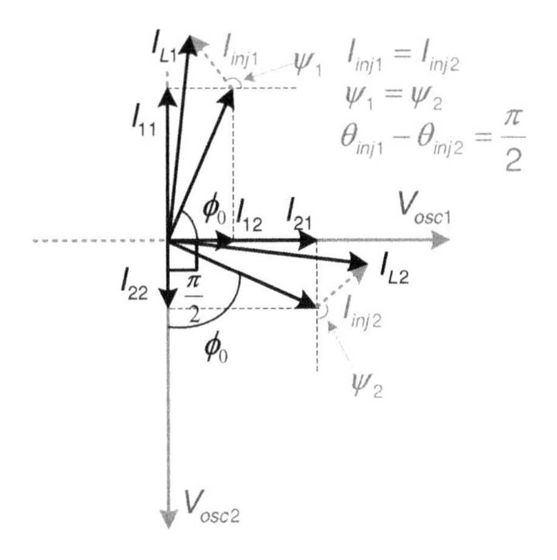

Fig. 4.12 Graphical representation of steady-state solution for the voltage and current phasors of the two-stage ring-oscillator when external signals are injected to both delay cells with the equal amplitudes, and quadrature phases

phase mismatch of

$$\theta_{\mathrm{mm}} \approx \frac{\pi}{4} \frac{I_{\mathrm{inj}}}{I_1^2} (I_1 \sin \psi + I_2 \cos \psi) \sin \varepsilon \tag{4.51}$$

where I_1 and I_2 are the bias currents of the negative resistance delay cell of Fig. 4.2b. In the derivation of (4.51), it is assumed that $\varepsilon \leq 10°$.

4.6 Phase Noise Analysis of Injection-Locked Regenerative Divider

To analyze the output phase noise of the injection-locked regenerative frequency divider, we use the simplified block diagram of this divider with the main phase noise sources shown in Fig. 4.13. It is assumed that the oscillator's internal phase noise and the phase noise of the input frequency to the divider are the main contributors to the output phase noise.

To start the analysis, we assume that steady-state is reached and the oscillator is super-harmonic injection-locked to the Rth harmonic of its output frequency.

It is shown in [19] that a super-harmonic injection-locked oscillator, to the Rth harmonic of its output frequency, behaves like a first-order PLL with an input phase to output phase transfer function of $G(S) = \frac{1}{1+S/\omega_P}$ where $S = j\Delta\omega$ and as shown in [1], ω_P for the injection-locked oscillator of Fig. 4.8 is approximately

$$\omega_P = \frac{\frac{\pi}{4} I_{\mathrm{inj}} \cos \psi}{I_1 + \frac{\pi}{4} I_{\mathrm{inj}} \sin \psi} \omega_{\mathrm{inj}}. \tag{4.52}$$

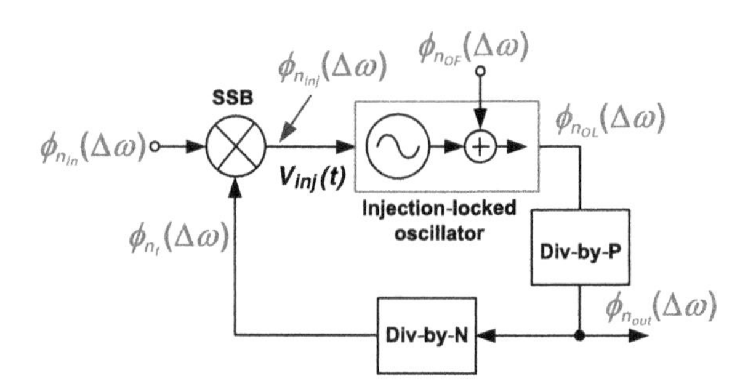

Fig. 4.13 Phase noise mechanism in an injection-locked regenerative frequency divider

Consequently, the phase noise at the output of the injection-locked oscillator ($\overline{\phi_{n\mathrm{OL}}^2}(\Delta\omega)$) can be expressed in terms of the oscillator's intrinsic noise ($\overline{\phi_{n\mathrm{OF}}^2}(\Delta\omega)$) and the phase noise of the injection signal ($\overline{\phi_{n\mathrm{inj}}^2}(\Delta\omega)$).

$$\overline{\phi_{n\mathrm{OL}}^2}(\Delta\omega) = \frac{1}{R^2}\frac{1}{1+(\Delta\omega/\omega_P)^2}\overline{\phi_{n\mathrm{inj}}^2}(\Delta\omega) + \frac{(\Delta\omega/\omega_P)^2}{1+(\Delta\omega/\omega_P)^2}\overline{\phi_{n\mathrm{OF}}^2}(\Delta\omega) \tag{4.53}$$

As can be seen from (4.53), the phase noise at the output of the oscillator consists of two components: the oscillator's intrinsic phase noise, which goes through a first-order high-pass transfer function, and the phase noise of the injection signal which encounters a first-order low-pass transfer function. Consequently, the close-in phase noise of the injection-locked oscillator is dominated by the phase noise of the injection signal, while the high-frequency phase noise follows the internal phase noise of the free-running oscillator. Therefore, $\overline{\phi_{n\mathrm{OL}}^2}(\Delta\omega)$ can be approximated for close-in phase ($\Delta\omega \ll \omega_P$) and high-frequency phase noise ($\Delta\omega \gg \omega_P$) by

$$\overline{\phi_{n\mathrm{OL}}^2}(\Delta\omega) \approx \begin{cases} \frac{1}{R^2}\overline{\phi_{n\mathrm{inj}}^2}(\Delta\omega) & \Delta\omega \ll \omega_P \\ \overline{\phi_{n\mathrm{OF}}^2}(\Delta\omega) & \Delta\omega \gg \omega_P \end{cases}. \tag{4.54}$$

As discussed in [20], if the SSB phase noise of the inputs to the mixer of Fig. 4.1c are $\overline{\phi_{n\mathrm{in}}^2}$ and $\overline{\phi_{nf}^2}$, and if the amplitude of the signal at the mixer output is sufficiently large, the output phase noise of the mixer is mainly dominated by the phase noise of the inputs and can be written as

$$\overline{\phi_{n\mathrm{inj}}^2}(\Delta\omega) = \overline{\phi_{n\mathrm{in}}^2}(\Delta\omega) + \overline{\phi_{nf}^2}(\Delta\omega). \tag{4.55}$$

Since the divide-by-P is locked to the oscillator output, we can assume that its output phase noise is dominated by the phase noise of the injection-locked oscillator, i.e. $\overline{\phi_{n\mathrm{out}}^2}(\Delta\omega) = \overline{\phi_{n\mathrm{OL}}^2}(\Delta\omega)/P^2$. As a result, the phase noise of the feedback signal (the output of the divide-by-N) is

$$\overline{\phi^2_{nf}}(\Delta\omega) = \overline{\phi^2_{n\mathrm{OL}}}(\Delta\omega)/(NP)^2. \tag{4.56}$$

Substituting (4.55) and (4.56) into (4.54), the output phase noise of the injection-locked regenerative divider, $\pounds_{\mathrm{out}}\{\Delta\omega\} = 10\log[\overline{\phi^2_{n\mathrm{out}}}(\Delta\omega)]$, can be expressed as

$$\pounds_{\mathrm{out}}\{\Delta\omega\} \approx \begin{cases} 10\log\left[\dfrac{\overline{\phi^2_{n\mathrm{in}}}(\Delta\omega)}{(RP)^2 - 1/N^2}\right] & \Delta\omega \ll \omega_P \\ 10\log\left[\dfrac{\overline{\phi^2_{n\mathrm{OF}}}(\Delta\omega)}{P^2}\right] & \Delta\omega \gg \omega_P \end{cases}. \tag{4.57}$$

It can be concluded from (4.53) that if the injection signal is noise-less, the phase noise at the output of the frequency divider would be equal to the attenuated extrinsic phase noise of the oscillator [15, 19], in other words

$$\pounds_{\mathrm{out}}\{\Delta\omega\} = 10\log\left[\frac{(\Delta\omega/\omega_P)^2}{1+(\Delta\omega/\omega_P)^2}\frac{\overline{\phi^2_{n\mathrm{OF}}}(\Delta\omega)}{P^2}\right]. \tag{4.58}$$

To complete this section, we investigate the phase noise of the two-stage ring-oscillator of Fig. 4.2a. The free-running phase noise of a two-stage ring-oscillator-based VCO that uses negative resistance delay cells with active loads is calculated in [21] using the impulse sensitivity function (ISF) technique of [22, 23]. Here we use the result obtained in [21] to find the phase noise of the two-stage ring-oscillator of Fig. 4.2. In this case, the free-running phase noise at the offset $\Delta\omega$ from the carrier (in dBc/Hz) can be expressed as

$$\pounds_{\mathrm{FR}}\{\Delta\omega\} = 10\log\left[\frac{\pi^2}{24}\frac{(\overline{i_n^2}/\Delta f)}{(CV_{\mathrm{sw}})^2(\Delta\omega)^2}\right] \tag{4.59}$$

where $\overline{i_n^2}/\Delta f$ is the output referred mean square current noise density which contains both thermal and flicker noises, and can be expressed by $\overline{i_n^2}(M1, M2, M3, M4)/\Delta f + 2\times 4\mathrm{kT}/R_L$. In this expression, R_L is the load resistor used in the delay cell of Fig. 4.2b since it is the only noisy component of the output resistance at the output of each delay cell. Using the oscillation frequency given by (4.25), the phase noise (4.59) can be re-written as

$$\pounds_{\mathrm{FR}}\{\Delta\omega\} = 10\log\left[\frac{\pi^2}{24}\left(\frac{RI_2}{V_{\mathrm{sw}}I_1}\right)^2\frac{(\overline{i_n^2}/\Delta f)}{(\Delta\omega/\omega_{\mathrm{SRF}})^2}\right]. \tag{4.60}$$

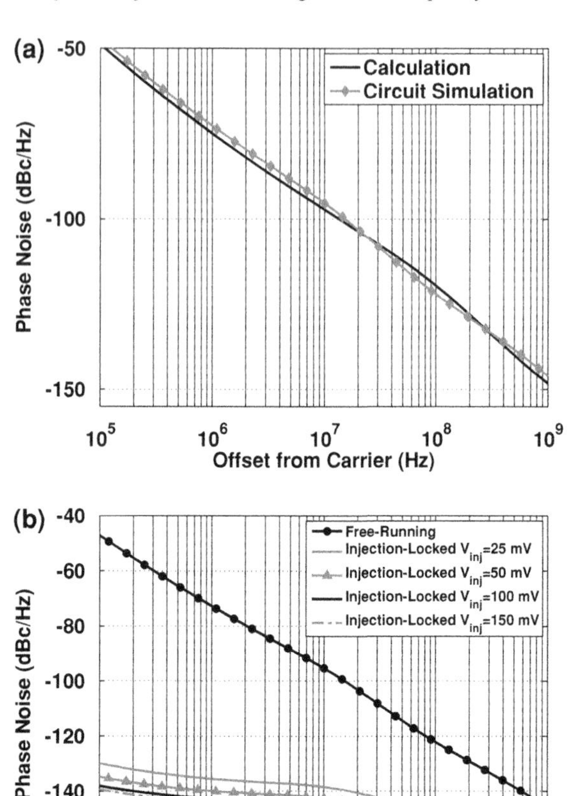

Fig. 4.14 Phase noise of the prototype two-stage ring-oscillator. **a** Calculated phase noise from (4.60) versus circuit simulation of free-running phase noise, and **b** circuit simulated phase noise when injection-locked to a noise-less injection signal for different values of injection signal (external signal is injected to both of the delay cells)

The simulated and calculated free-running phase noise of the prototype two-stage ring-oscillator from Sect. 4.5.2 is plotted in Fig. 4.14a. To precisely calculate the phase noise, $\overline{i_n^2}/\Delta f$ is measured using circuit simulation, and then substituted into (4.60). The simulated injection-locked phase noise of this two-stage ring-oscillator for different amplitudes of the injection signal is shown in Fig. 4.14b. To obtain these plots, this two-stage ring-oscillator is used as a divide-by-two and noise-less differential injection signals are applied to the gates of transistors labeled by $MT1$ (Fig. 4.2b) in each delay cell. As was discussed before, the output phase noise of the oscillator is its attenuated internal noise, and this noise is more attenuated for a larger amplitudes of injection signal.

Using (4.57), (4.60), and the phase noise of the injection signal, one can calculate the output phase noise of the injection-locked regenerative frequency divider.

4.7 Design Example of a Divide-by-2.25/4.5

As mentioned earlier, one of the benefits of the injection-locked regenerative frequency divider is its ability to obtain fractional division ratios while providing 50 % duty cycle quadrature output phases. In this section we present a design example of an injection-locked regenerative frequency divider that generates a fractional division ratio.

One of the design goals is to achieve a design suitable for implementation in a digital CMOS submicron technology using no on-chip inductor. This goal implies using ring-oscillators with resistive loads and also exploiting SSB mixers in the injection-locked regenerative frequency divider. The latter requires availability of quadrature phases of both the input to the frequency divider, and its output, or more precisely, the output of the feedback path.

There are several ways to provide quadrature phases of input, such as using polyphase filters, or preceding the divide-by-2.25 by another frequency divider or a ring-oscillator that can generate quadrature output phases. As previously stated, a block with quadrature outputs in the forward path is required to generate quadrature output phases. A divide-by-two flip-flop is usually a good candidate for this purpose, since most divide-by-two circuits can provide quadrature output phases. On the other hand, any oscillator that is injection-locked to its second harmonic can be used as a divide-by-two. Two-stage, or four-stage, ring-oscillators are good examples that serve as divide-by-two circuits and provide quadrature outputs.

Figure 4.15 shows a fractional frequency divider based on the general architecture of Fig. 4.1c. This frequency divider is primarily designed to obtain a division ratio of 2.25 to be used for frequency synthesis for Multiband OFDM UWB [1]. The injection-locked frequency divider (ILFD) in the forward path of the frequency divider of Fig. 4.15 is implemented using the two-stage ring-oscillator with negative resistance delay cells, shown in Fig. 4.2. In addition to operating with a smaller input drive, an ILFD can operate at higher speeds in comparison to static frequency dividers. The feedback path of the frequency divider of Fig. 4.15 consists of a cascade of two flip-flop-based CML divide-by-two circuits. Using divide-by-two blocks in the feedback path provides quadrature phases for the operation of the SSB mixer.

From (4.1) it can be concluded that in the steady-state, $f_{out1} = f_{in}/2.25$ and $f_{out2} = f_{in}/4.5$. Therefore, the frequency divider of Fig. 4.15 achieves division ratios of both 2.25 and 4.5. In contrast to the previous approaches [3], no on-chip inductor is required to implement this frequency divider.

As mentioned in Sect. 4.5.3, when appropriate injection signals are applied to both the delay cells of the ring-oscillator of Fig. 4.2, there is no systematic quadrature mismatch at the output. Monte-Carlo simulation using the extracted layout of the frequency divider was performed to investigate the quadrature output mismatches induced by process variations, device mismatches, and layout imperfections. The simulation results showed a standard deviation of 0.3 dB for quadrature amplitude mismatch, and a standard deviation of 1.4° for quadrature phase error across the locking range. These amplitude and phase mismatches are tolerable for most

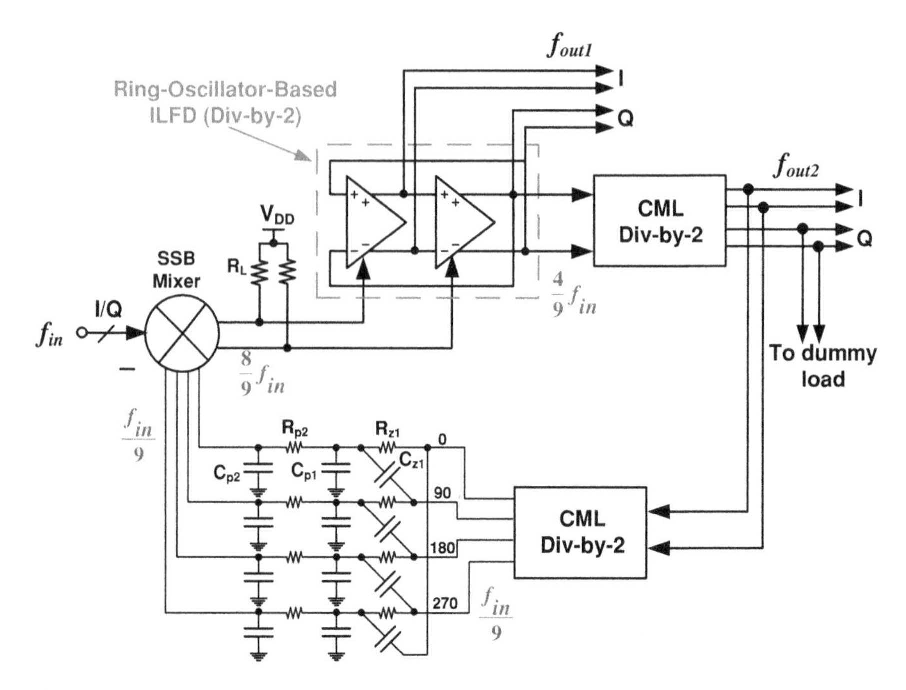

Fig. 4.15 Simplified schematic of a divide-by-2.25/4.5 with quadrature outputs and 50% duty cycle using an injection-locked regenerative frequency divider architecture. A single-sideband mixer based on Gilbert cell mixer with resistive load is used. Dummy loads at the output of frequency dividers are not shown

applications. The frequency divider of Fig. 4.15 and its output buffers (not shown in Fig. 4.15), draw 14 mA from a 1.2 V supply.

The frequency divider of Fig. 4.15 is similar to the divide-by-4.5 presented in [24] where the forward path consists of a cascade of two divide-by-two, and the feedback path is implemented using a divide-by-two. The first divide-by-two in the forward path is a CML divider, and the other two frequency dividers are implemented using true single-phase clocked (TSPC) logic [25] to reduce the power dissipation. The SSB mixer in [24] uses an inductive load to suppress unwanted mixing products. The frequency divider of [24] does not use any linearization technique or harmonics polyphase filtering used in the frequency divider of Fig. 4.15.

The input signal to the frequency divider can be applied to either the RF port (Gm stage) or the LO port of the SSB mixer. When it is connected to the RF port, as shown in Fig. 4.16a, the divider achieves a better input sensitivity; thus, it can function with smaller input power. On the other hand, the output of the feedback path frequency divider is fed to the LO port. The large swing of the CML divider output is suitable for saturation operation of the LO port of the mixer. However, in this case, all the odd harmonics of the feedback signal contribute to in-band mixing products at the mixer output since the feedback signal is at a lower frequency than the input signal. Consequently, the signal at the output of the mixer can achieve a

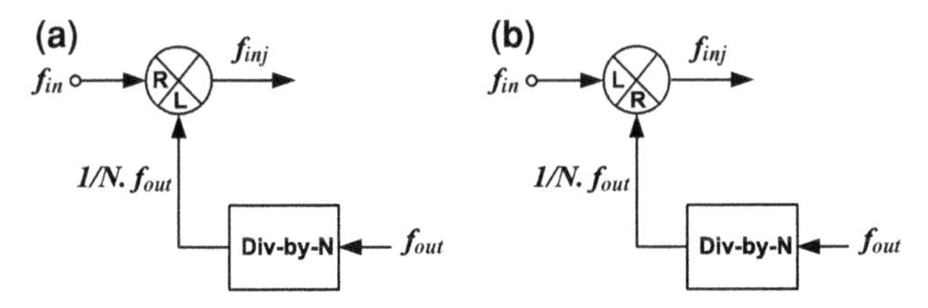

Fig. 4.16 Different options to implement the frequency divider of Fig. 4.15. **a** input signal to the frequency divider is applied to the RF port (Gm stage) of the SSB mixer, **b** input signal to the frequency divider is applied to the LO port (switches) of the SSB mixer

non-50 % duty cycle which leads to the I/Q phase inaccuracy at the final output of the main frequency divider. Moreover, as shown in [26], any frequency spur at the input of a divide-by-two translates to a spur at the output of the divider at the same offset frequency.

To solve these issues, one can use the scheme shown in Fig. 4.16b in which the input signal, which is at a greater frequency than the feedback signal, is applied to the LO port of the SSB mixer. In this scheme the feedback signal is applied to the RF port (Gm stage) of the mixer. The contribution of the feedback signal to higher-order in-band mixing products can be minimized by linearizing the Gm stage of the mixer. In addition, the feedback signal, which is rich in harmonics, can go through harmonic suppression filtering. In the presence of the quadrature phases of the feedback signal, a polyphase filter can be used since: (1) It provides balanced loading for all the outputs of the feedback frequency divider. (2) It can achieve better harmonic suppression by generating imaginary zeros. (3) It does not require any on-chip inductor or balun. The drawback of the scheme shown in Fig. 4.16b is the degraded input sensitivity of the resultant frequency divider.

In this scheme, if the input signal is smaller than required for the saturation operation of the mixer LO port, the amplitude of the mixer output signal depends on the input amplitude, and hence could be small. As previously stated, using an ILFD in the forward path of the divide-by-2.25 has the advantage that it can operate with smaller input drive. As a result, it can guarantee robust operation of the injection-locked regenerative divider of Fig. 4.15.

As was discussed in Sect. 4.4, the choice of this ILFD, or generally the forward path oscillator, is directly tied to the stability of the injection-locked regenerative divider. Nonetheless, using an ILFD mandates careful analysis and simulations to make sure that it has a wide enough locking range to compensate for process variations. A ring-oscillator based ILFD has a wider locking range compared to LC oscillator. In addition, it can provide multiple phases of output, occupies smaller area on silicon, and is more compatible with the digital CMOS technology.

As was shown in Sect. 4.4, the locking range of the injection-locked regenerative divider follows the locking range of its oscillator. When the schemes of Fig. 4.16b i used to implement the frequency divider of Fig. 4.15, the mixer (Fig. 4.15) is set by the output of its preceding frequency divider. So, in order to control the amplitude of

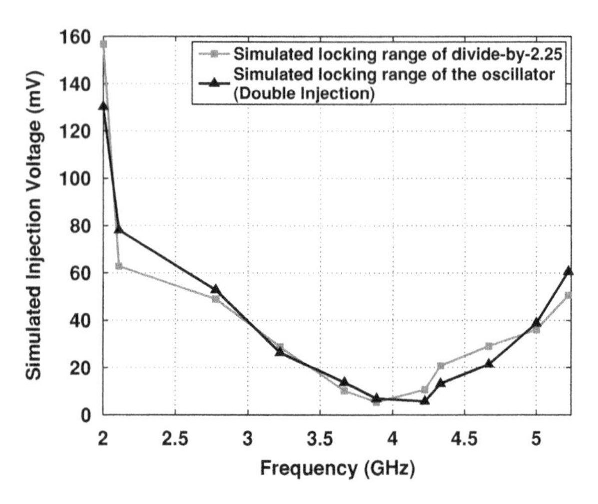

Fig. 4.17 Circuit simulated locking range of the prototype divide-by-2.25/4.5, and comparison with the locking range of the oscillator used in it

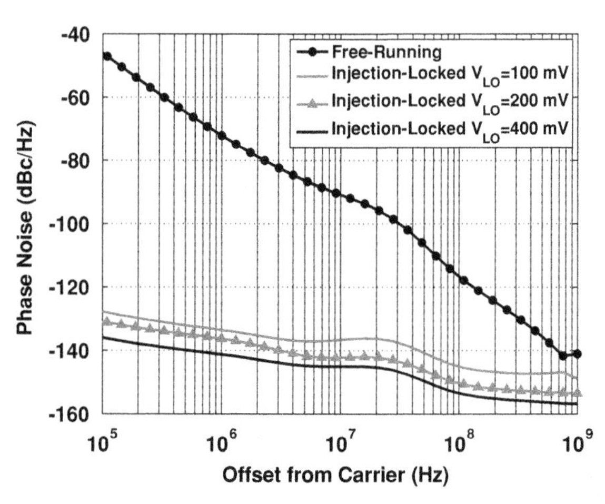

Fig. 4.18 Circuit simulation of free-running and injection-locked phase noise of the prototype divide-by-2.25/4.5 for different LO amplitudes

$V_{\text{inj}}(t)$, the amplitude of the input to the LO port of the mixer is varied. The simulated locking range of this divider is shown in Fig. 4.17 and is compared with the locking range of the two-stage negative resistance-based ring-oscillator. As can be seen, the agreement is very good throughout the locking range and the maximum error in predicting the injection voltage is about 30 mV at the low end of the locking range.

Figure 4.18 shows the simulated free-running and locked phase noise of the frequency divider for an output frequency of 4 GHz, when a noise-less input signal is input to the divider. The amplitude of the injection signal is changed by changing the amplitude of the input to the LO port of the mixer. From (4.58) we expect to obtain similar output phase noise to those in Fig. 4.14, and comparing Fig. 4.18 and 4.14 shows a very good agreement within 2 dB.

4.8 Conclusion

This chapter presented an analysis of the operation, stability, locking range, and phase noise of injection-locked regenerative frequency dividers. In addition, the injection-locked behavior of two-stage ring-oscillators (based on negative resistance delay cells) is studied and their locking range is derived for the first time. Finally, a design technique was presented for implementing a regenerative frequency divider in a digital CMOS technology (using no on-chip inductor or balun) for achieving fractional division ratios with a 50 % duty cycle quadrature output phases. The circuit simulation results of the designed oscillator and the fractional injection-locked regenerative frequency divider are in excellent agreement with the calculations.

References

1. Farazian M (2009) Fast hopping high-frequency carrier generation in digital CMOS technology. Ph.D. dissertation, University of California, San Diego, May 2009
2. Farazian M, Gudem P, Larson L (2010) Fast hopping carrier generation for 14- band multi-band OFDM UWB in digital CMOS. Topical meeting on silicon monolithic integrated circuits in RF systems, Jan 2010
3. Lin C, Wang C-K (2005) A regenerative semi-dynamic frequency divider for mode-1 MB-OFDM UWB hopping carrier generation. In: IEEE ISSCC digest of technical papers, vol 1, pp 206–207, Feb 2005
4. Razavi B (2003) Phase-locking in high-performance systems: from devices to architectures. Wiley, New York
5. Miller R (1939) Fractional-frequency generators utilizing regenerative modulation. Proc IRE 27(7):446–457
6. Safarian A, Anand S, Heydari P (2006) On the dynamics of regenerative frequency dividers. IEEE Trans Circuits Syst II: Express Briefs 53(12):1413–1417
7. Sengupta K, Bhattacharyya T, Hashemi H (2007) A nonlinear transient analysis of regenerative frequency dividers. IEEE Trans Circuits Syst I: Regul Pap 54(12):2646–2660
8. Zheng H, Luong H (2008) A double-balanced quadrature-input quadrature-output regenerative frequency divider for UWB synthesizer applications. IEEE Trans Circuits Syst I: Regul Pap 55(9):2944–2951
9. Li J-L, Qu S-W, Xue Q (2008) A theoretical and experimental study of injection-locked fractional frequency dividers. IEEE Trans Microw Theory Tech 56(11):2399–2408
10. Razavi B (2007) Heterodyne phase locking: a technique for high-frequency division. In: IEEE ISSCC digest of technical papers, pp 428–429, Feb 2007
11. Larsson P (2000) Fractional frequency divider. US Patent 6,157,694, Dec 2000
12. Razavi B (2004) A study of injection locking and pulling in oscillators. IEEE J Solid-State Circuits 39(9):1415–1424
13. Farazian M, Gudem P, Larson L (2009) A CMOS multi-phase injection-locked frequency divider for V-band operation. IEEE Microw Wireless Compon Lett 3:447–450
14. Adler R (June 1946) A study of locking phenomena in oscillators. Proc IRE 34(6):351–357
15. Mirzaei A, Heidari M, Bagheri R, Abidi A (2008) Multi-phase injection widens lock range of ring-oscillator-based frequency dividers. IEEE J Solid-State Circuits 43(3):656–671
16. Chien J, Lu L (2007) Analysis and design of wideband injection-locked ring-oscillators with multip-le-input injection. IEEE J Solid-State Circuits 42(9):1906–1915
17. Gangasani G, Kinget P (2008) Time-domain model for injection locking in nonharmonic oscillators. IEEE Trans Circuits Syst I: Regul Pap 55(6):1648–1658

18. Wan Y, Lai X, Roychowdhury J (2005) Understanding injection locking in negative-resistance LC oscillators intuitively using nonlinear feedback analysis. In: Proceedings of the IEEE custom integrated circuits conference, pp 729–732, Sept 2005
19. Verma S, Rategh H, Lee T (2003) A unified model for injection-locked frequency dividers. IEEE J Solid-State Circuits 38(6):1015–1027
20. Mishra C, Valdes-Garcia A, Bahmani F, Batra A, Sanchez-Sinencio E, Silva-Martinez J (Dec. 2005) Frequency planning and synthesizer architectures for multiband OFDM UWB radios. IEEE Trans Microw Theory Tech 53(12):3744–3756
21. Yan W, Luong H (2001) A 900-MHz CMOS low-phase-noise voltage-controlled ring-oscillator. IEEE Trans Circuits Syst II: Analog Digit Signal Process 48(2):216–221
22. Hajimiri A, Lee T (1998) A general theory of phase noise in electrical oscillators. IEEE J Solid-State Circuits 33(2):179–194
23. Hajimiri A, Limotyrakis S, Lee T (1998) Phase noise in multi-gigahertz CMOS ring-oscillators. In: Proceedings of the IEEE custom integrated circuits conference, pp 49–52, May 1998
24. Kuo Y, Weng R (2008) Regenerative frequency divider for 14 sub-band UWB applications. Electron Lett 44(2):111–112
25. Rabaey J, Chandrakasan A, Nikolic B (2002) Digital integrated circuits. Prentice Hall, Englewood Cliffs
26. Ismail A, Abidi A (2005) A 3.1- to 8.2-GHz zero-IF receiver and direct frequency synthesizer in 0.18μm SiGe BiCMOS for mode-2 MB-OFDM UWB communication. IEEE J Solid-State Circuits 40(12):2573–2582

Chapter 5
Design of Broadband Amplifiers in Digital CMOS Technology

5.1 Overview

The frequency synthesizer of Fig. 2.24 uses a cascade of two single-sideband mixers to implement all fourteen center frequencies of MB-OFDM UWB. This frequency synthesizer has a similar structure to the signal path of a wideband transmitter, and therefore broadband amplifiers maybe required to boost the signal at different places. An optimized combination of mixers and broadband amplifiers is needed to implement this wideband frequency synthesizer, and can help to reduce the level of in-band spurious tones. In this chapter, we study different options to implement a cascade of two mixers, as well as the interface circuitry needed to implement this cascade of mixers.

5.2 Architectures for Cascading Two Mixers

An illustration of a series connection of two frequency mixers is shown in Fig. 5.1. The fundamental frequency components at the output of $Mixer$ 2 in Fig. 5.1 are as follows:

$$f_{\mathrm{out}} = |f_{LO2} \pm (f_{LO1} \pm f_{RF1})|. \tag{5.1}$$

One can select a single frequency component from (5.1) by using single-sideband mixers for $Mixer$ 1 and $Mixer$ 2 and appropriate quadrature phases for $RF1$, $LO1$, and $LO2$.

As can be seen from Fig. 5.1, the output of the first mixer is used as an input to the second mixer. The output signal of $Mixer$ 1 can be either current or voltage.The following Sections explain these two approaches in more detail, and also compare them against each other.

M. Farazian et al., *Fast Hopping Frequency Generation in Digital CMOS*,
DOI: 10.1007/978-1-4614-0490-3_5,

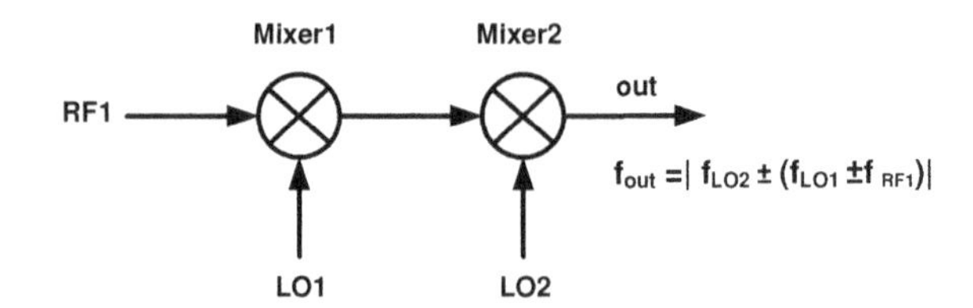

Fig. 5.1 An illustration of cascading two frequency mixers

5.2.1 Voltage-Mode Approaches for Cascading of Mixers

In this case, the output voltage of the first mixer is the input to the transconductor or switching core of the second mixer. Two examples of this approach are shown in Fig. 5.2a, b. In Fig. 5.2a, the output current of the first mixer is converted to voltage in the resistor load, and then it is fed to the transconductor stage of the second mixer. The architecture of Fig. 5.2b, on the other hand, connects the output voltage of the first mixer to the switching core of the second mixer.

If the resistor load of the first mixer is replaced by a tuned LC circuit, the output can generate adequate voltage swing to drive the transconductor or the switching core of the second mixer. However, in order to comply with an inductor-less design methodology the mixers used here only use resistive loads. The voltage swing at the output of the first mixer becomes more important in Fig. 5.2b where the output voltage of the first mixer goes to the switching core of the second mixer. In this case, a larger voltage swing is required to achieve complete switching operation, and avoid degradation of the conversion gain and output noise of the second mixer. These issues can be solved by inserting an amplifier at the output of the first mixer to bring the signal to an adequate level for the second mixer, as shown in Fig. 5.3.

In summary, the voltage-mode approach shown in Fig. 5.1 is straightforward, but it relies on utilizing a high-gain broadband amplifier to drive the capacitive load of the gates of the FET switches of the second mixer. As the last note on this architecture, the frequency synthesizer of Fig. 2.24 uses single-sideband mixers, and also implements both in-phase and quadrature output phases of the output signals. So, four single-sideband mixers (two single-sideband mixers for each stage) and two interface amplifiers are required. The use of two single-sideband mixers for the second stage of mixing increases the capacitive load of the interface amplifier.

5.2.2 Current-Mode Approaches for Cascading of Mixers

Another approach to the implementation of the cascade of two mixers shown in Fig. 5.1 is to use the output current of the first mixer. This scheme eliminates the need for current-to-voltage conversion at the output of the first mixer, and also eliminates the transconductance stage of the second mixer. A simplified block diagram of this technique is shown in Fig. 5.4.

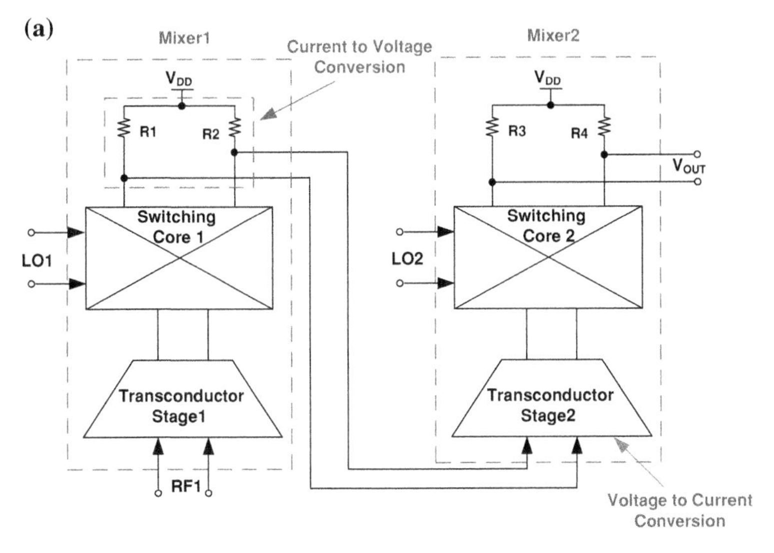

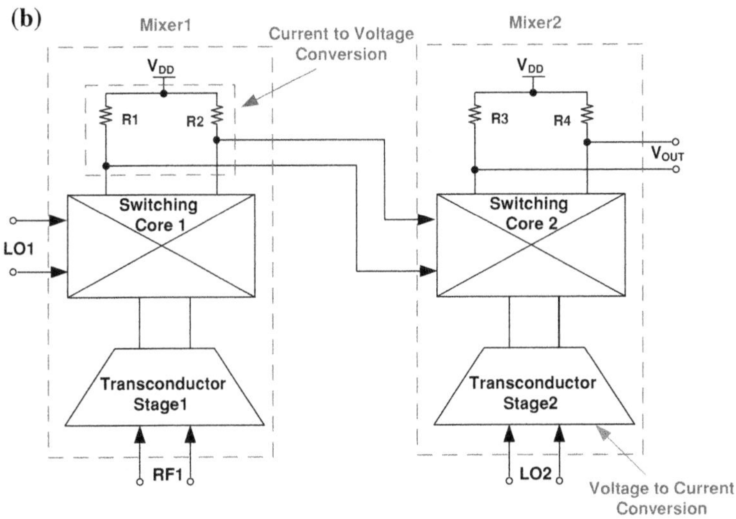

Fig. 5.2 Architectures to implement a voltage-mode cascade of two mixers, **a** by applying the output voltage of the first mixer to the transconductor of the second mixer, **b** by applying the output voltage of the first mixer to the switching core of the second mixer

5.2.2.1 Examples of Current-Mode Implementation for Cascading of Mixers

Figure 5.5 shows an example of a current-mode implementation of the cascade of two double-balanced mixers.The input of the switching core, which is the source of an NMOS or a PMOS transistor, is a low impedance node. Consequently, the output of the first mixer in Fig. 5.5 (the common drain of $M3$ and $M5$ or $M4$ and

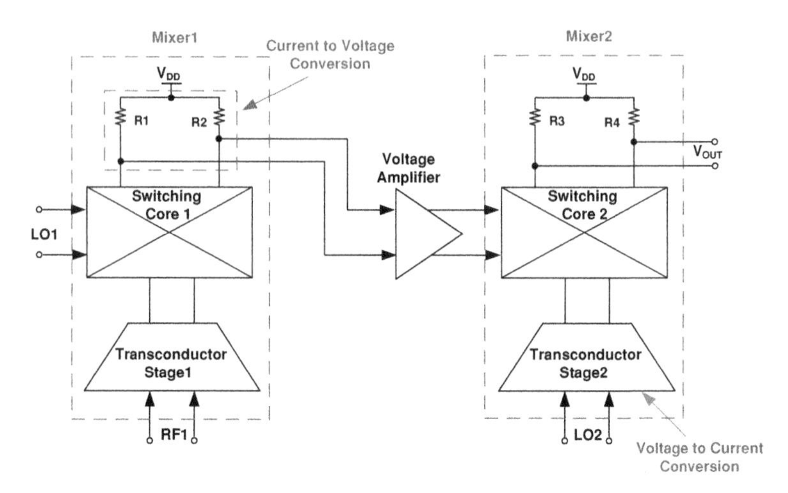

Fig. 5.3 Insertion of amplifier at output of first mixer

*M*6) has an improved frequency response compared to its corresponding node in the architectures of Fig. 5.2a, b. As mentioned earlier, the frequency synthesizer of Fig. 2.24 uses single-sideband mixers, and also implements both in-phase and quadrature output phases of the output signals. This implies replacing the double-balanced mixer of Fig. 5.5 with a single-sideband mixer. In addition, single-sideband operation is enabled in the second switching core by inserting cascode transistors ($MC1$, $MC2$, $MC3$, and $MC4$), as shown in Fig. 5.6. Transistors $MC1$, $MC2$, $MC3$, and $MC4$ split the output current of the first switching core and enable using that current in a single-sideband mixer (both in-phase and quadrature parts of a single-sideband mixer). Figure 5.6 only shows the half circuit and the other half (for quadrature output) is not shown.

The architectures of Figs. 5.5 and 5.6 require a stack of four and five transistors, respectively. Hence, these architectures are not suitable for low-voltage implementation, and they require a voltage supply higher than 1.2 V to be implemented in a standard 0.13 μm technology.

One alternative for the double-balanced mixer of Fig. 5.5 that overcomes the headroom issues is shown in Fig. 5.7. In the double-balanced mixer of Fig. 5.7, the output current of the first mixer is mirrored and fed to another switching core (the PMOS switching core in this example). Converting the double-balanced mixer of Fig. 5.7 to a single-sideband mixer does not pose any headroom issues and can be achieved by adding two more PMOS transistors to copy the currents of $MP2$ and $MP3$.

The mixer of Fig. 5.7 solves the headroom issues of the mixer shown in Fig. 5.5 by adding the $MP2$ and $MP3$ diode-connected devices at the output of the first mixer and using $MP1$ and $MP4$ as current mirrors to provide the second switching core by the output current of the first mixer. However, the addition of these PMOS devices increases the capacitive load at the output of the first mixer and lowers the frequency of the pole that is formed at this node. In addition, large PMOS devices

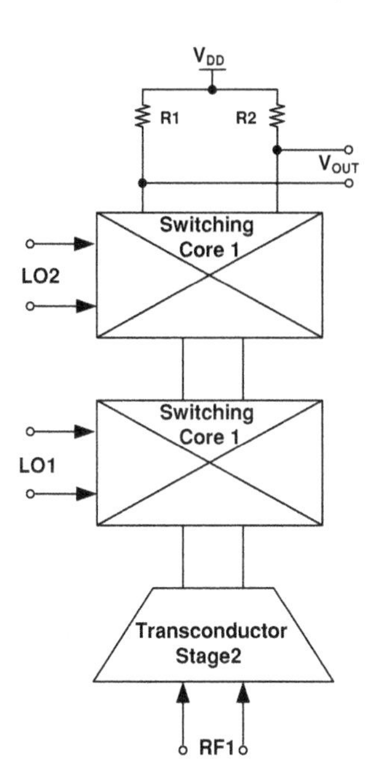

Fig. 5.4 A current-mode cascade of two mixers

are required to reduce the required headroom for transistors $MP1$–$MP4$ and enable a low-voltage implementation for the architecture of Fig. 5.7. But using large PMOS devices is at the price of increasing the capacitive load at the output of the first mixer.

If the transconductor and the first switching core in the architecture of Fig. 5.7 (transistors $M1$–$M6$) are implemented using PMOS devices, then NMOS diode-connected devices and current mirrors are required to feed the output current of the fist single-sideband mixer to the second switching core. Usage of NMOS devices adds less parasitic capacitance to the output of the first switching core; hence, it improves the frequency response by increasing the pole frequency formed at this node. However, this approach requires larger devices in the transconductor ($M1$ and $M2$) to achieve the same transconductance (g_m) and conversion gain for the same bias current.

As will be discussed in Chap. 6 the transconductor in the single-sideband mixer needs to be linearized in order to mitigate the spurious tones at the output. Adding a linearization scheme to the transconductor may require additional voltage headroom, and it will make it more difficult to implement the stacked architectures of Fig. 5.5 and Fig. 5.7 with a voltage supply of less than 1.2 V. More details on the linearization of the transconductance stage of the mixer will be discussed in Chap. 6.

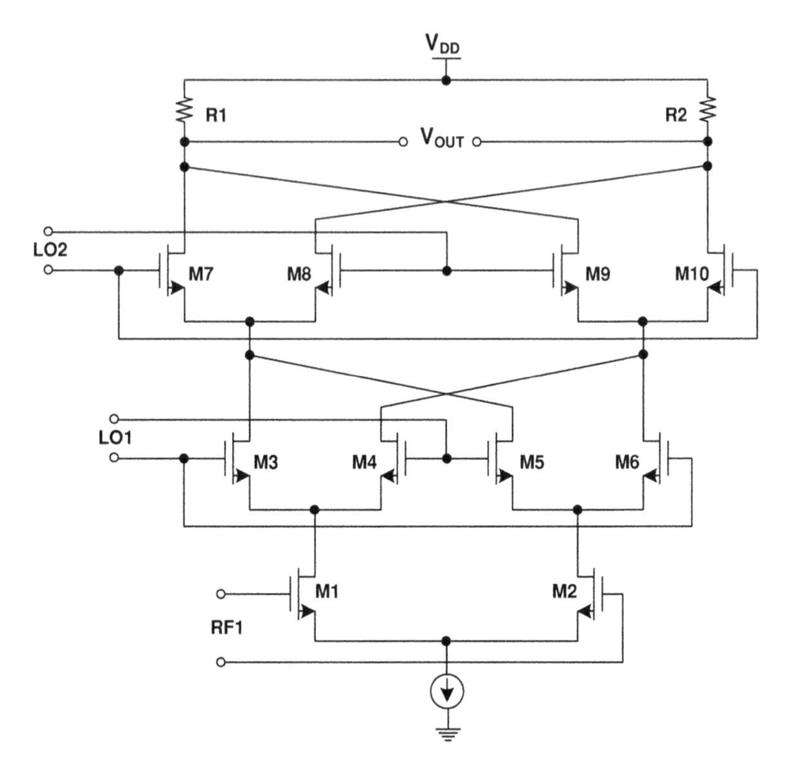

Fig. 5.5 Two-stage stacked double-balanced mixer

5.2.3 Comparison of Mixer Cascading Topologies

The stacked mixer configuration of Fig. 5.4 is the simplest way to implement a cascade of two mixers with no additional interface amplifier, but it requires a stack of four or five transistors as shown in Figs. 5.5, 5.6, and so is not well suited for a low-voltage CMOS implementation. The architecture shown in Fig. 5.7 is a modified version that works with a lower voltage supply. However, there is a significant loss of signal at the output of the first mixer due to the large capacitive load of the current mirrors. Consequently, current-mode solutions for implementing a cascade of two mixers are problematic at a low-voltage supply.

On the other hand, the voltage-mode approach of Fig. 5.2 for cascading two single-sideband mixers is more suitable for a low-voltage implementation, but it relies on the availability of a high-gain broadband amplifier. Figure 5.8 shows the half circuit of the block diagram for voltage-mode implementation for cascading two single-sideband mixers.

The gain-bandwidth requirements of the broadband interface amplifier of Fig. 5.8 is similar to the gain-bandwidth requirements of the limiting amplifiers that are used in optical communications [1–3]. Consequently, one can employ the design

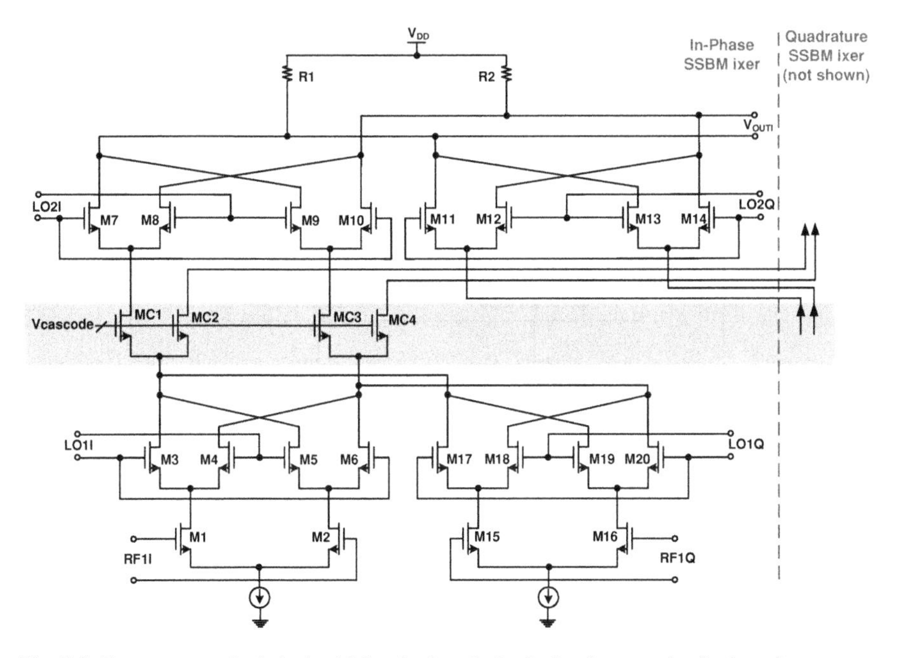

Fig. 5.6 Two-stage stacked single-sideband mixer (only the in-phase section is shown)

techniques used for the design of limiting amplifiers to design the broadband interface amplifier of Fig. 5.8. Limiting amplifiers are used to boost the output of a front-end transimpedance amplifier (TIA) and bring it to the appropriate levels that are suitable for the operation of digital logic. While limiting amplifiers provide a large gain and operate at high frequencies up to several GHz, they usually have to drive a 50-ohm load [4], whereas in this case capacitive loads predominate. The rest of this chapter continues with a brief overview of broadband amplifier design techniques that meets the required gain-bandwidth product and can be implemented in digital CMOS technology.

5.3 Overview of Design Techniques for CMOS Broadband Amplifiers

5.3.1 Cascade of Low-Gain Wideband Amplifiers

Achievement of a large voltage gain in a single-stage amplifier relies on cascoding to increase the output impedance, which is not an attractive option for low-voltage design since it requires additional voltage headroom for the cascode transistors. Consequently, a multi-stage amplifier approach is preferred. The drawback of using

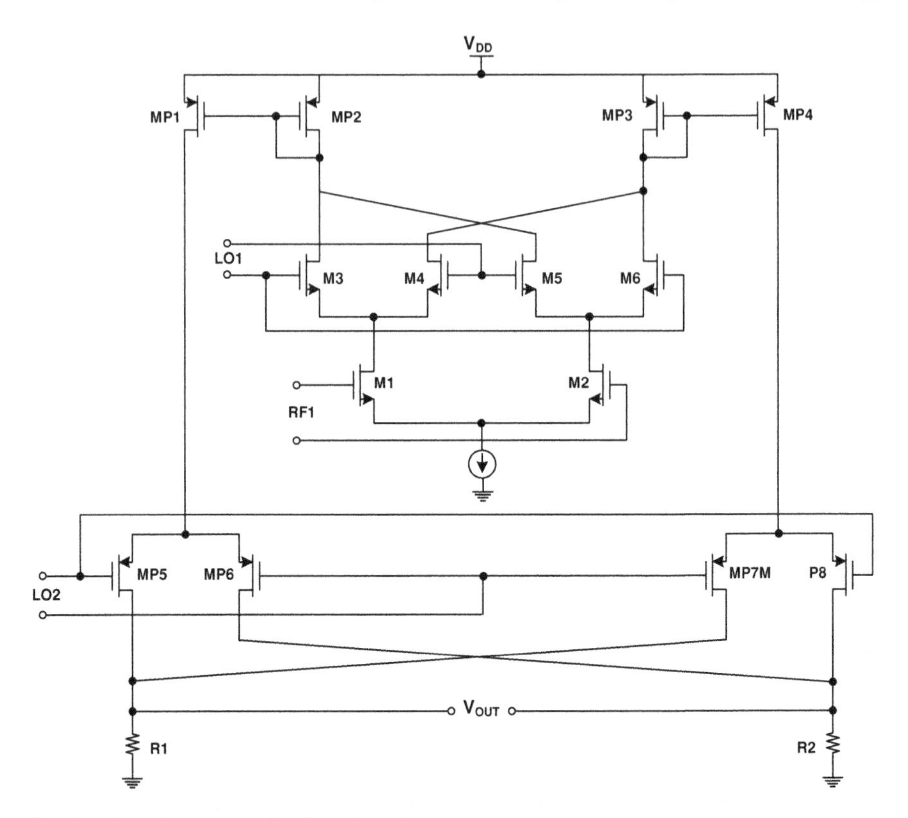

Fig. 5.7 Folded two-stage double-balanced mixer

a multi-stage amplifier is that each amplifier introduces an additional pole that limits the achievable bandwidth. In addition, the Miller effect leads to an increase in the input capacitive load of each amplifier and consequently an increase in the capacitive load of the preceding stage. Therefore, a low-voltage gain is desired, but it is at the price of increasing the number of gain stages in the cascade.

One common technique to overcome this issue is to use a cascade of low-gain single-pole wideband amplifiers. Cascading N identical single-pole amplifiers with a DC voltage gain of A_o and a 3-dB bandwidth of ω_o results in a composite amplifier with a voltage gain of

$$A(s) = \left(\frac{A_o}{1 + \frac{s}{\omega_o}} \right)^N . \tag{5.2}$$

Using (5.2), The 3-dB bandwidth of this composite amplifier (ω_1) is

$$\omega_1 = \omega_o \sqrt{2^{1/N} - 1}. \tag{5.3}$$

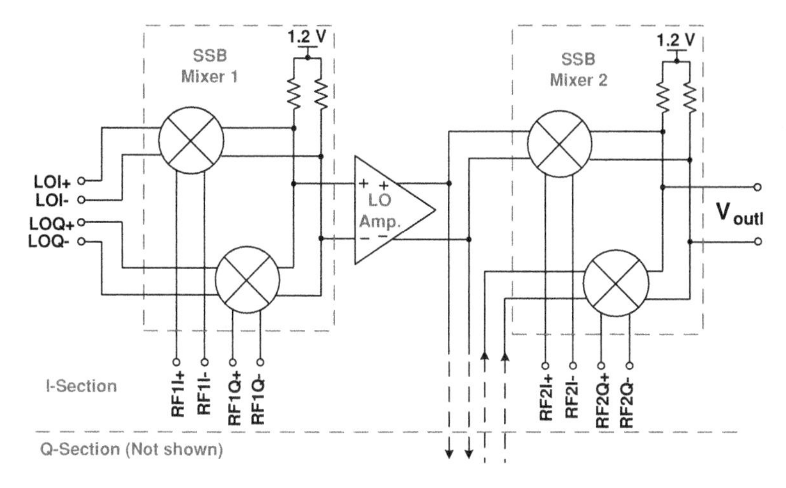

Fig. 5.8 Mixing stages and the amplifier interface (only the in-phase section is shown)

Different values for N result in different values for gain per stage and the 3-dB bandwidth in the single-pole wideband amplifiers when one tries to to meet a target voltage gain of A_T. Assuming that the unity gain bandwidth of each single-pole wideband amplifier is fixed and is given by $A_o\omega_o$, a larger value for N results in a smaller gain per stage and a larger 3-dB bandwidth in each single-pole amplifier. One performance metric in this method is to maximize the ratio of the 3-dB bandwidth of the overall amplifier to the unity gain bandwidth of each amplifier, i.e., $\omega_1/(A_o\omega_o)$ [4, 5]. However, as discussed in [4], there is no significant improvement in the ratio of $\omega_1/(A_o\omega_o)$ when the number of stages goes beyond five, while at the same time increasing the number of stages reduces the gain per stage and increases the noise contribution of each stage.

5.3.2 Neutralization for Bandwidth Enhancement

Neutralization is a commonly used technique to reduce the input capacitance of an amplifier. The principles of neutralization are very similar to the Miller effect. The Miller effect leads to an increase in the input capacitance of an inverting voltage-mode amplifier due to any parasitic capacitance between its input and output nodes, but in neutralization one uses an auxiliary capacitor between the same amplifier input node and uses a non-inverting gain, that is equal to the magnitude of the inverting gain of the amplifier. It can be shown that this leads to the cancelation of the Miller effect [6]. It is much easier to implement neutralization in an amplifier with differential outputs since both paths are available. Figure 5.9 shows an example of neutralization in a CMOS fully differential amplifier.

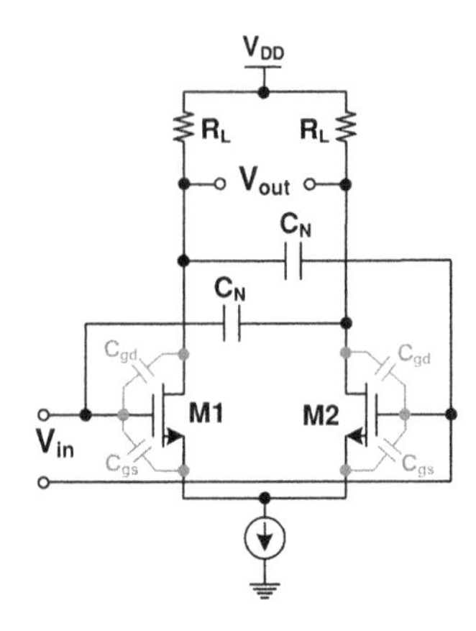

Fig. 5.9 Example of neutralization in a fully differential amplifier

In the fully differential amplifier of Fig. 5.9, the capacitance at the gate of transistor $M1$ can be expressed by

$$C_{\text{in}} \cong C_{\text{gs}} + C_{\text{gd}} + C_{\text{N}} + |A_o|(C_{\text{gd}} - C_{\text{N}}) \tag{5.4}$$

where C_{gs} and C_{gd} are the intrinsic gate–source and gate–drain capacitances of transistor $M1$, respectively, and C_{N} is the neutralization capacitor. In (5.4), A_o is the voltage gain from the gate to the drain of $M1$, and can be expressed in terms of the transconductance and the output resistance of $M1$ by $-g_{m1}(r_{\text{ds1}}||R_L)$. The single-ended output capacitance of the amplifier of Fig. 5.9 caused by C_{gd} and C_{N} is

$$C_{\text{out}} \cong C_{gd} + C_{\text{N}} + \frac{C_{gd} - C_{\text{N}}}{|A_o|}. \tag{5.5}$$

In neutralization, C_{N} is usually chosen to be equal to C_{gd} to cancel the Miller effect. However, in this case both input and output capacitances of the amplifier of Fig. 5.9 have a residual value of $C_{gd} + C_{\text{N}}$. One can choose C_{N} to completely cancel the C_{gd} component from the input capacitance of the amplifier of Fig. 5.9. In this case, the single-ended input capacitance of the amplifier of Fig. 5.9 is C_{gs} and its single-ended output capacitance is

$$C_{\text{out}} = 2C_{gd}\left(1 + \frac{1}{|A_o|}\right). \tag{5.6}$$

One problem in neutralization is controlling C_{N} to track C_{gd}. Neutralization does not completely cancel the gate–drain capacitor component (C_{gd}) from the input impedance of the amplifier of Fig. 5.9 if the neutralization capacitor (C_{N}) is smaller than the gate–drain capacitor. On the other hand, a larger neutralization capacitor than the gate–drain capacitor may lead to an inductive component in the input impedance of this amplifier that causes peaking in the frequency response. This inductive

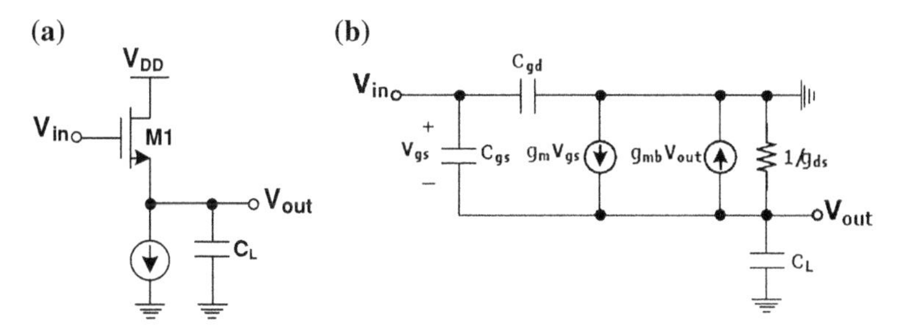

Fig. 5.10 **a** A source follower amplifier (buffer), and **b** the small-signal model of the source follower amplifier.

component in the input impedance can cause instability in a high-gain amplifier or in a cascade of amplifiers.

5.3.3 Unity Gain Buffers

Another technique that facilitates the design of high-gain multi-stage broadband amplifiers is inserting a unity gain voltage buffer between an amplifier and its load. This technique preserves the voltage gain and extends the bandwidth of the amplifier by reducing its capacitive load.

A source follower can be used as a unity gain buffer. A CMOS source follower buffer and its small-signal model are shown in Fig. 5.10a, b, respectively. In the small-signal model of Fig. 5.10b, g_m is the transconductance, g_{mb} represents the body effect, and $1/g_{ds}$ is the finite drain–source resistance of transistor $M1$.

The voltage gain of the source follower of Fig. 5.10a can be derived from its small-signal model.

$$\frac{V_{\text{out}}(s)}{V_{\text{in}}(s)} = \frac{g_m + sC_{gs}}{g_m + g_{\text{mb}} + g_{\text{ds}} + sC_{gs} + sC_{\text{L}}}. \tag{5.7}$$

From (5.7), the low-frequency gain of the source follower of Fig. 5.10a is less than unity mainly due to the body effect and also due to the finite output resistance of $M1$. The body effect and the finite output impedance of transistor $M1$ also affect the input capacitance of the source follower. The low-frequency input capacitance of the source follower of Fig. 5.10a is

$$C_{\text{in}} = C_{gd} + \frac{g_{\text{mb}} + g_{\text{ds}}}{g_m + g_{\text{mb}} + g_{\text{ds}}} C_{gs}. \tag{5.8}$$

The body effect and the finite output resistance lead to a degradation of the voltage gain and an increase in the input capacitance as shown by (5.7) and (5.8). In addition,

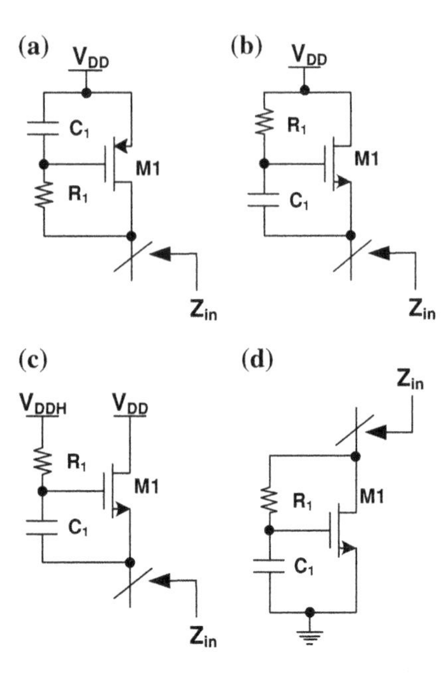

Fig. 5.11 Different schemes to implement an active inductor, **a** using a PMOS diode-connected transistor, **b** using an NMOS transistor , **c** modified version of **b** for low-voltage operation, and **d** using an NMOS diode-connected device

as discussed in [7] one cannot simultaneously obtain a wide bandwidth and a large capacitance driving capability by inserting buffers at the output of each gain stage. Therefore, only one driver stage is usually used at the output of the amplifier to drive the large output capacitive load. This will be discussed in more details in Sect. 5.3.6.

5.3.4 Shunt Peaking and Active Inductors

Shunt peaking is one way to expand the bandwidth of amplifiers by adding an inductive load [5, 8]. In this method, the inductive load moves the poles of the amplifier to higher frequencies by canceling some portion of the capacitive load of the amplifier. Hence, it can increase the bandwidth of an amplifier by up to 70 % depending on the value of the inductor [8].

One can use active inductors in the absence of passive on-chip inductors for shunt peaking and bandwidth expansion. There are a variety of architectures that can implement an active inductor in CMOS technology [2, 9–12].

An example of an active inductor is shown in Fig. 5.11a. This circuit consists of a PMOS diode-connected transistor ($M1$), a resistor (R_1), and a capacitor (C_1). The circuit can be implemented using an NMOS transistor to reduce the parasitic capacitances of $M1$, as shown in Fig. 5.11b.

The input impedance of the active inductors of Fig. 5.11a, b is

$$Z_{\text{in}}(s) = \frac{1}{g_m + g_{\text{ds}}} \times \frac{1 + s/\omega_z}{1 + s/\omega_p} \tag{5.9}$$

where g_m and g_{ds} represent the transconductance and the finite output resistance ($1/r_{\text{ds}}$) of transistor $M1$. In 5.9, ω_z and ω_p can be expressed as

$$\omega_z = \frac{1}{R_1(C_1 + C_{\text{gs}})} \tag{5.10a}$$

$$\omega_p = \frac{g_m + g_{\text{ds}}}{(1 + g_{\text{ds}} R_1)(C_1 + C_{\text{gs}})}. \tag{5.10b}$$

If $g_m R_1$ is greater than unity, then ω_p is greater than ω_z and the input impedances of the circuits shown in Fig. 5.11a or b have inductive behavior.

The maximum voltage at the input of Fig. 5.11a, b must be less than $V_{\text{DD}} - |V_{\text{GS}}|$. Consequently, these active inductors are not a proper choice for low-voltage operation. The circuit shown in Fig. 5.11c is a slightly modified version of Fig. 5.11b that can achieve a higher voltage swing, depending on the gate bias voltage (V_{DDH}) [2]. Lastly, the active inductor of Fig. 5.11d is similar to the one shown in Fig. 5.11b. However, it is equivalent to an inductor with one terminal connected to ground.

To summarize, all the designs in Fig. 5.11 require a $g_m R_1 > 1$ in order to provide inductive input impedance. This condition may not be true at a large voltage swing, where g_m is reduced. In addition, the structures of Fig. 5.11b and d require a supply voltage of higher than 1.2 V for proper operation.

5.3.5 Cherry–Hooper Amplifier

One of the problems of implementing a high-gain broadband amplifier with a resistive load is the well-known trade-off between the gain and frequency response. Achieving a high gain at a low supply voltage usually requires a multi-stage amplifier with large input devices to achieve large transconductance and sufficiently large load resistors for each stage to meet the required voltage gain and voltage swing for the multi-stage amplifier.

The Cherry–Hooper amplifier is a two-stage common source amplifier that uses feedback to reduce the input and output resistances of its second stage. Hence, it pushes the dominant poles to higher frequencies and achieves a wider bandwidth [13]. A standard implementation of a fully differential Cherry–Hooper amplifier is shown in Fig. 5.12a [4, 14]. In this implementation, the feedback resistor across the input transistors of the second stage ($M3$ and $M4$) help to lower the input and

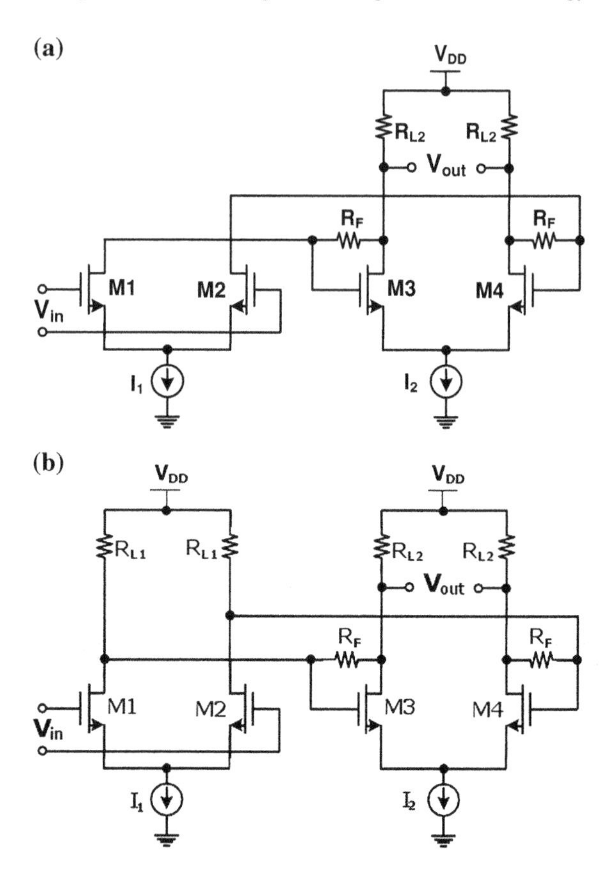

Fig. 5.12 Fully different Cherry–Hooper amplifier, **a** standard implementation **b** modified implementation with resistive load

output resistances of the second stage. Assuming $g_{m3}R_F \gg 1$, the single-ended input resistance at the gate of $M3$ (or $M4$) can be expressed as

$$R_{in} \approx \frac{R_F}{1 + g_{m3}(R_F||R_{L2}||r_{ds3})} \tag{5.11}$$

where g_{m3} and r_{ds3} are the transconductance and the finite output resistance of transistor $M3$ or $M4$. Similarly, the single-ended output resistance of the second stage seen at the drain of $M3$ (or $M4$) is as follows.

$$R_{out} = \frac{1}{g_{m3}}||R_{L2}||r_{ds3} \tag{5.12}$$

These single-ended input and output resistances both simplify to $1/g_{m3}$ if $R_F \ll R_{L2}, r_{ds3}$. Consequently, the following conditions must hold in order to have a single-ended input and output resistance of $1/g_{m3}$ at the input and the output of the second stage of the amplifier of Fig. 5.12a.

$$R_{L2}, r_{ds3} \gg R_F \gg \frac{1}{g_{m3}} \tag{5.13}$$

Under these conditions, the low-frequency voltage gain of the Cherry–Hooper amplifier of Fig. 5.12a is

$$\frac{V_{out}}{V_{in}} \approx g_{m3} R_F. \tag{5.14}$$

The conditions stated by 5.13 are not easy to meet at a low supply voltage since a large g_{m3} is required to push the output poles of the first and the second stages to high frequencies and achieve a wide bandwidth. A large g_{m3} is achieved by using a combination of a large transistor size for $M3$ and $M4$ and also a large bias current for these transistors (I_2). The former adds more parasitic capacitances to the output of the first stage and lowers the frequency of the pole at that node while the latter requires more headroom across R_{L2}. In addition, the bias current of the first stage flows through R_F and R_{L2}. This demands a higher supply voltage to design the Cherry–Hooper amplifier of Fig. 5.12a [4]. Hence, this architecture is not suitable for a low supply voltage implementation.

One alternative that can reduce the DC voltage drop across R_{L2} is shown in Fig. 5.12b [4]. Here, the load resistors of the first stage provides a portion of the bias current. This reduces the current flowing through R_{L2}, and leads to a lower required supply voltage for the operation of the second stage. The voltage gain of the amplifier of Fig. 5.12b is as described by (5.14) if $R_{L1} \gg R_F$. A smaller value for R_{L1} leads to the degradation of the voltage gain of the amplifier of Fig. 5.12b. However, this new condition and the conditions of (5.13) cannot be met at a low supply voltage.

Lastly, a source follower buffer (or a common emitter buffer in a BiCMOS technology) can be used to implement the feedback from the drain to the gate of $M3$ (and $M4$) and provide a small input resistance at the output of the first stage as described in [1, 4, 14, 15]. This way the total bias current of the first stage and the second stage are drawn from the power supply separately and the voltage drop across the load resistor of the second stage is significantly reduced compared to the standard implementation of a Cherry–Hooper amplifier shown in Fig. 5.12a. However, usage of a CMOS source follower puts more constraints on the minimum supply voltage required to implement this variation. In summary, the Cherry–Hooper amplifier loses its advantages at a low supply voltage, i.e., it is very difficult to achieve a large gain and a wide bandwidth.

5.3.6 Driver and Output Stages

A driver stage is used at the output of high-speed CMOS amplifiers to drive the capacitive load to prevent the degradation of the frequency response [2, 3]. This is very similar to using unity gain buffers discussed in Sect. 5.3.3. However, unity gain buffers are used between different stages in a multi-stage amplifier while driver and

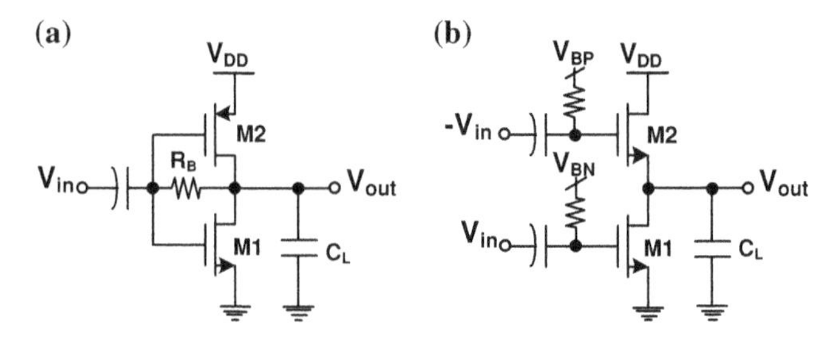

Fig. 5.13 Different options for buffers in digital CMOS technology, **a** inverter-based buffer, **b** NMOS only buffer

output stages are used between the amplifier and its load. In other words, the output stages must be able to drive a larger capacitive load.

The driver stage usually handles larger voltage swings compared to the input of an amplifier, so a large-signal analysis is considered.

The source follower of Fig. 5.10a can be used as a driver stage and shows a small output impedance, and its voltage gain at high frequencies between its pole frequency and zero frequency ($g_m/(C_{gs}+C_L) < \omega < g_m/C_{gs}$) can be approximated by g_m/sC_L. In large voltage operation, $M1$ is off during the negative voltage swing. Hence, the discharge rate of the load capacitance during the negative input voltage swing is limited by the current source used in the source follower of Fig. 5.10a.

One alternative for this buffer is the inverter-based configuration of Fig. 5.13a. In spite of its larger output impedance ($1/(g_{ds1} + g_{ds2})$) compared to the source follower, the inverter-based buffer achieves a small-signal voltage gain of $(g_{m1} + g_{m2})/sC_L$ at high frequencies, which is larger than the small-signal voltage gain of a source follower buffer. In addition, the charging rate of the load capacitance is now determined by $M2$ and the discharging rate is set by $M1$. Hence, it can serve as a better driver stage. But the PMOS transistor needs to be larger than the NMOS transistor to achieve the same current handling that the NMOS transistor achieves, and this adds more parasitic capacitance to the input and the output.

In a fully differential system, one can swap the PMOS transistor of an inverter-based buffer with an NMOS transistor that is driven with the inverted phase of the input signal as shown in Fig. 5.13b. This buffer consists of an NMOS common source amplifier and an NMOS source follower, and it has a smaller input and output parasitic capacitance compared to the inverter-based buffer of Fig. 5.13a. In this buffer, two NMOS transistors are AC coupled to different phases of the input signal and are biased independently. Two different bias voltages can be chosen to achieve the best linearity and the driving strength. If we ignore the gate–source and the gate–drain capacitances of transistors $M1$ and $M2$, the small-signal voltage gain of the buffer of Fig. 5.13b can be expressed by

$$\frac{V_{out}(s)}{V_{in}(s)} = \frac{g_{m1} + g_{m2}}{g_{m2} + g_{mb2} + g_{ds1} + g_{ds2} + sC_L}. \tag{5.15}$$

As can be seen from (5.15), the high frequency small-signal gain of this buffer is approximately $(g_{m1} + g_{m2})/sC_L$ which is similar to the gain of the inverter-based buffer. In spite of the inverter-based buffer of Fig. 5.13a, the buffer of Fig. 5.13b provides a low output impedance even at low frequencies, i.e.,

$$Z_{out}(s) = \frac{1}{g_{m2} + g_{mb2} + g_{ds1} + g_{ds2} + sC_L}. \quad (5.16)$$

One can design the buffer of Fig. 5.13b to operate with the same DC bias voltage for transistors $M1$ and $M2$. This can be done by using NMOS transistors with different threshold voltages, i.e., by using a standard threshold voltage transistor for $M1$, and a low-threshold voltage NMOS transistor for $M2$. As a result V_{BP} and V_{BN} are equal, and at least one of the AC coupling capacitors in the buffer of Fig. 5.13b can be eliminated. The elimination of AC coupling capacitors saves some chip area, and also reduces the parasitic capacitances introduced by the AC coupling capacitors. The driver stages shown in [16, 17] use an NMOS transistor with a low-threshold voltage for the source follower and an NMOS with a high-threshold voltage for the common source amplifier of the driver stage of Fig. 5.13b.

To summarize, the buffer of Fig. 5.13a is suitable to be used as a high frequency driver stage to drive a capacitive load due to its low output impedance, smaller input and output parasitic capacitances compared to an inverter-based buffer, and its large voltage gain.

5.4 Proposed Inductor-Less Broadband Amplifier

In this section, we present our proposed inductor-less broadband amplifier. This broadband amplifier is shown in Fig. 5.14a and uses a combination of the broadband design techniques presented in Sect. 5.3. Our proposed amplifier consists of three broadband low-gain single-stage amplifiers (Amp1, Amp2, and Amp3) and one driver stage to drive the gates of the FET switches. The single-stage amplifier of Fig. 5.14b is used for Amp1, Amp2, and Amp3. The combination of cross-coupled transistors $M2$ and $M4$ and the capacitor C_N in Fig. 5.14b is used in [3] as a negative capacitance cell to enhance the bandwidth of an amplifier by canceling some portion of its output load or parasitic capacitances. A further detailed analysis is done in [18] and it is shown that this cell represents a negative resistance in parallel with a frequency dependent negative capacitance.

A very comprehensive analysis of the small-signal and large-signal output admittance of Fig. 5.14b is done in [19] and the details are presented in Appendix C of this book. As shown in [19], The small-signal output admittance of this circuit is

$$Y_N(s) = 2sC_{gd2} - sC_N \frac{g_{m2} - sC_{gs2}}{g_{m2} + s(C_{gs2} + 2C_N)} \quad (5.17)$$

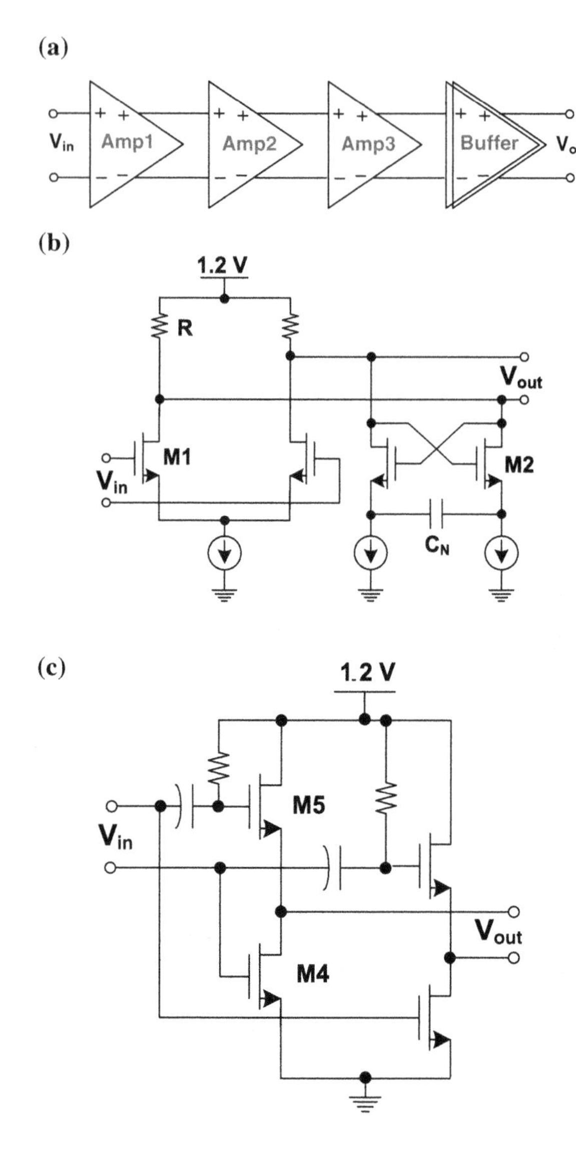

Fig. 5.14 Proposed broadband amplifier and its building blocks, **a** top level schematic, **b** resistive load differential amplifier with negative impedance generator, and **c** output driver stage

where C_{gs2}, C_{gd2}, and g_{m2} are the gate–source capacitor, gate–drain capacitor, and the transconductance of the cross-coupled transistors ($M2$ and $M4$), respectively. The large-signal output admittance of this amplifier is analyzed in [19], and is presented in Appendix C of this book.

In the broadband amplifier of Fig. 5.14a, Amp1, Amp2, and Amp3 are designed to provide enough amplification for the output of the first single-sideband mixer to drive the switches of the second mixer. As discussed in Sect. 5.3.6, an output driver stage is used after the last amplifier in the cascade to drive the capacitive load of the

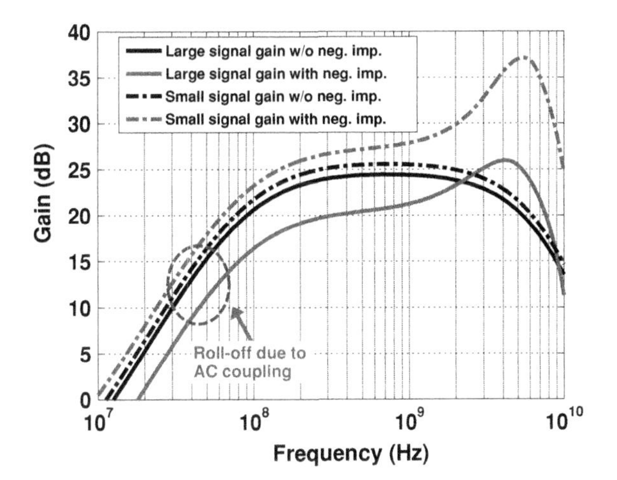

Fig. 5.15 Simulated small-signal and large-signal gain of the LO amplifier of the Fig. 5.14

gates of FET switches of the switches of the second single-sideband mixer without degradation of the bandwidth of the overall amplifier. The schematic of this output driver is shown in Fig. 5.14c. This driver stage is the differential version of the NMOS buffer of Fig. 5.13b.

Figure 5.15 shows the simulated voltage gain of the amplifier of Fig. 5.14a and shows a comparison between resistive loads and the negative impedance generator of Fig. 5.14b. Both small-signal and large-signal simulation results are presented for these configurations.

Figure 5.15 shows a larger difference between the small-signal and the large-signal simulated voltage gain of the amplifier of Fig. 5.14b when the resistive load differential amplifier with negative impedance generator of Fig. 5.15 is used to implement Amp1, Amp2, and Amp3 compared to the case where Amp1, Amp2, and Amp3 are implemented using a simple differential amplifier with resistive load. However, the negative impedance generator leads to gain peaking and bandwidth expansion at the frequency of interest, even in a large-signal regime.

References

1. Holdenried C, Haslett J, Lynch M (2004) Analysis and design of HBT Cherry- Hooper amplifiers with emitter-follower feedback for optical communications. IEEE J Solid-State Circuits 39(11):1959–1967
2. Sackinger E, Fischer W (2000) A 3 GHz, 32 dB CMOS limiting amplifier for SONET OC-48 receivers. In: IEEE ISSCC digest of technical papers, pp 158–159, Feb 2000
3. Galal S, Razavi B (2003) 10-Gb/s limiting amplifier and laser/modulator driver in 0.18 μm CMOS technology. IEEE J Solid-State Circuits 38(12):2138–2146
4. Razavi B (2003) Design of integrated circuits for optical communications. McGraw- Hill, New York

5. Lee T (2004) The design of CMOS radio-frequency integrated circuits. Cambridge University Press, Cambridge
6. Kreithen A (1951) Neutralization of amplifiers. US Patent 2,542,087, 20 Feb 1951
7. Sackinger E, Fischer W (2000) A 3 GHz 32 dB CMOS limiting amplifier for sonnet oc-48 receivers. IEEE J Solid-State Circuits 35(12):1884–888
8. Mohan S, Hershenson M, Boyd S, Lee T (2000) Bandwidth extension in CMOS with optimized on-chip inductors. IEEE J Solid-State Circuits 35(3):346–355
9. Hara S, Tokumitsu T, Tanaka T, Aikawa M (1988) Broad-band monolithic microwave active inductor and its application tominiaturized wide-band amplifiers. IEEE Trans Microw Theory Tech 3(12):1920–1924
10. Thanachayanont A, Payne A (1996) VHF CMOS integrated active inductor. Electron Lett 32(11):999–1000
11. Hsiao C, Kuo C, Ho C, Chan Y (2002) Improved quality-factor of 0.18 μm CMOS active inductor by a feedback resistance design. IEEE Microwave Wirel Compon Lett 12(12):467–469
12. Thanachayanont A, Payne A (2000) CMOS floating active inductor and its applications to bandpass filter and oscillator designs. IEEE Proc Circuits Devices Syst 147(1):42–48
13. Cherry E, Hooper D (1963) The design of wide-band transistor feedback amplifiers. IEEE Proc 44(2):375–389
14. Holdenried C, Lynch M, Haslett J (2003) Modified CMOS cherry-hooper amplifiers with source follower feedback in 0.35 μm, technology. In: 29th european solid-state circuit conference, pp 553–556, Sept 2003
15. Abbott J, Plett C, Rogers J (2005) The design of wide-band transistor feedback amplifiers. In: Proceedings of the IEEE custom integrated circuits conference, 2005
16. von Buren G, Kromer C, Ellinger F, Huber A, Schmatz M, Jackel H (2006) A combined dynamic and static frequency divider for a 40 GHz PLL in 80 nm CMOS. IEEE ISSCC digest of technical papers, pp 2462–2471, Feb 2006
17. Kromer C, Sialm G, Berger C, Morf T, Schmatz M, Ellinger F, Erni D, Bona G-L, Jackel H (2005) A 100 mW 4×10 Gb/s transceiver in 80 nm CMOS for high-density optical interconnects. IEEE J Solid-State Circuits 40(12):2667–2679
18. Fanori L, Liscidini A, Catello R (2010) 3.3 GHz DCO with a frequency resolution of 150 Hz for all-digital PLL. In: IEEE ISSCC digest of technical papers, pp 48–50, Feb 2010
19. Farazian M (2009) Fast hopping high-frequency carrier generation in digital CMOS technology. Dissertation, University of California

Chapter 6
An Inductor-Less CMOS 14-Band Frequency Synthesizer for UWB

6.1 Overview

As mentioned in Chap. 1 one of the challenges in implementing a frequency synthesizer for Multi-band OFDM (MB-OFDM) UWB standard is overcoming the agility limitations of conventional synthesizers. The MB-OFDM proposal for UWB divides the spectrum from 3.1 to 10.6 GHz into 14 different bands, and frequency hops at the rate of 3.2 MHz between them [1] with a settling time of only 9.5 nS. The EVM requirements also pose challenging constraints on the spurious performance. Design techniques that eliminate the use of on-chip inductors, and which are compatible with low-voltage operation, are critical for increasing the level of integration.

An inductor-less design methodology may have several advantages over traditional design techniques: (1) While the area required to implement an on-chip inductor does not scale down in the finer technology nodes, inductor-less designs benefit from technology scaling. (2) On the other hand, the quality factor of the on-chip inductors may worsen in finer technology nodes, which can lead to an increase in the required current consumption to generate a given voltage swing. (3) It is more straightforward to port an inductor-less design into a new technology node. The penalty for an inductor-less design methodology is a slight increase in the current consumption to achieve the necessary gain and voltage swing in the absence of inductors.

In this work, appropriate frequency planning, harmonic suppression polyphase filters, and low-voltage linearization techniques in the single-sideband mixers (SSB) allowed us to remove the on-chip inductors in the synthesizer, while achieving sidebands as low as −38 dBc. This approach reduces die area and improves integrability with digital circuits, but may increase the power consumption. So low-power design techniques are essential.

The proposed architecture for a 14-band fast agile frequency synthesizer for UWB was discussed in Sect. 2.3, and it is redrawn in Fig. 6.1. The circuit design details for the implementation of this frequency synthesizer are discussed in Sect. 6.2, and

M. Farazian et al., *Fast Hopping Frequency Generation in Digital CMOS*,

DOI: 10.1007/978-1-4614-0490-3_6,

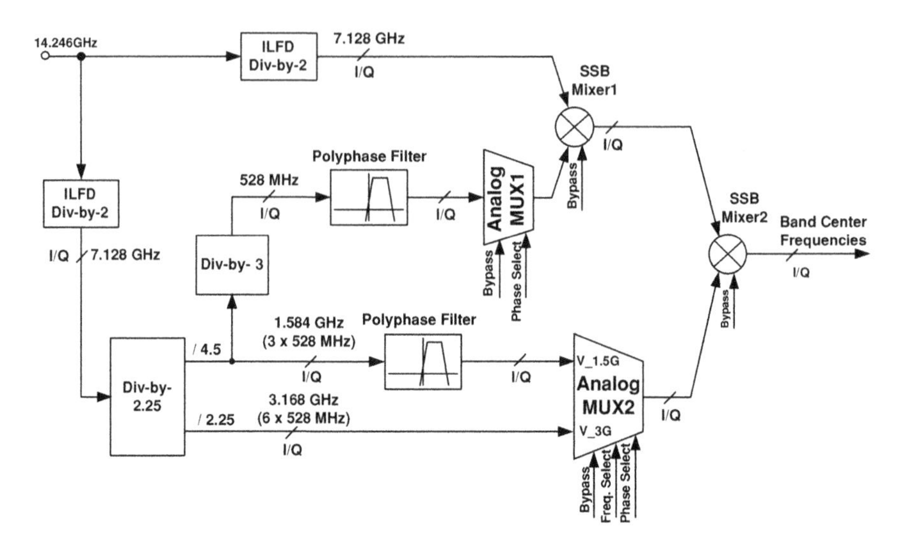

Fig. 6.1 Architecture for the universal 14-band UWB frequency synthesizer (redrawn from Chap. 2)

an inductor-less CMOS implementation of this frequency synthesizer along with the measured results of the silicon prototype are presented in Sect. 6.3.

6.2 Circuit Design

In this work, a combination of single-sideband mixers, polyphase filters, and appropriate linearization techniques help to meet the spurious requirements in the absence of on-chip inductors. A combination of different amplifier topologies is used to achieve the required voltage gain for interfacing between different blocks. Low-voltage design techniques are employed to make this implementation feasible at 1.2 V supply. In this section, circuit implementation details of the key blocks are presented.

6.2.1 Frequency Dividers

As was discussed in Sect. 2.3, implementation of the frequency synthesizer of Fig. 2.23 requires 528 MHz, 1.584, 3.168, and 7.128 GHz signals. The cascade of frequency dividers shown in Fig. 6.1 suggests a hardware efficient way to obtain all these required frequencies from a 14.256 GHz source. But these frequency dividers must also provide quadrature output phases. In addition, proper operation of the single-sideband mixers, with adequate LO leakage suppression requires input mixing signals with 50% duty cycle. It is straightforward to implement a

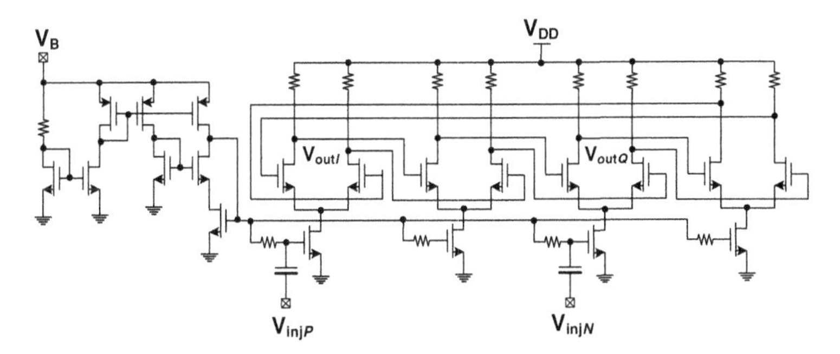

Fig. 6.2 Schematic of the four-stage ring-oscillator-based ILFD implemented using resistive load differential pair delay cell

divide-by-two with quadrature outputs and a 50 % duty cycle at the output. However, divide-by-three circuits often have a non-50 % duty cycle at the output, and achieving quadrature output phases is difficult [2, 3]. Lastly, frequency dividers with fractional division ratios (2.25 and 4.5) with 50 % duty cycle and quadrature outputs are required.

The front-end divide-by-two circuit in Fig. 6.1 is implemented using a four-stage ring-oscillator-based injection-locked frequency divider (ILFD) shown in Fig. 3.3. There are at least two advantages in using ILFDs instead of static frequency dividers: (1) The ILFDs operate at higher frequencies compared to static frequency dividers. (2) ILFDs require significantly smaller input drive for proper operation. The speed criteria changes for different CMOS technologies since static frequency dividers can achieve faster speeds in finer technology nodes. However, the required input drive of ILFDs is lower than that of static frequency dividers. This becomes more important at higher frequencies, where achieving large voltage signal swing is difficult. An inductor-less design methodology makes it even harder to achieve adequate input drive for operation of the static frequency dividers. This is the primary motivation for choosing an ILFD for the front-end frequency divider.

As shown in Chap. 3, the four-stage ring-oscillator-based frequency divider of Fig. 6.2 can achieve a division ratio of two at an input frequency of approximately twice its self-resonance frequency (SRF) when a sufficiently strong amplitude is applied. In this work, a single phase 14.256 GHz input is applied to the tail current source of the first delay stage. This frequency divider provides quadrature phases at 7.128 GHz, which will be used by the single-sideband mixer, and other frequency dividers. It was shown in Chap. 3 that when the frequency divider of Fig. 6.2 operates as a divide-by-two, it can achieve a frequency locking range of greater than 25 % around its SRF, which is greater than the simulated variation of its SRF at different process corners. Therefore, a wide locking range guarantees the proper operation of the ILFD after fabrication.

Regenerative frequency dividers [4] are a technique to obtain a fractional division ratios. However, as was discussed in Chap. 4, regenerative frequency dividers usually require some tuned circuit in the forward path. Moreover, they usually do not provide

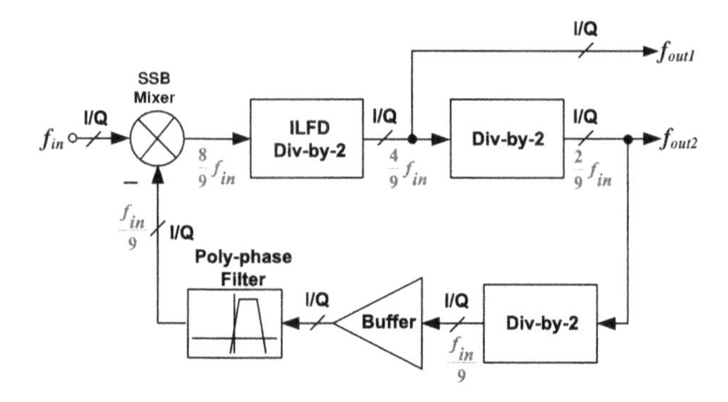

Fig. 6.3 Block diagram of the injection-locked regenerative divide-by-2.25 with quadrature outputs

quadrature output phases [4]. In this work, in order to achieve both fractional division ratio and quadrature output phases, an injection-locked regenerative frequency divider is used.

Figure 4.15 shows our proposed divide-by-2.25 circuit. This frequency divider is designed using the procedure described in Sect. 4.7 and it operates at an input frequency of 7.128 GHz, achieves division ratios of both 2.25 and 4.5, and provides quadrature outputs with a 50 % duty cycle. In contrast to the previous approaches [5], this frequency divider is implemented using no on-chip inductor.

An ILFD divide-by-two, implemented using a two-stage ring-oscillator (with negative resistance stages), is used in the forward path, and a cascade of two CML divide-by-two circuits is used in the feedback path. A mixer is needed for frequency translation and to close the feedback loop. A single-sideband mixer is used to generate the difference of the input frequency (f_{in}) and the output of the feedback dividers ($f_{\text{out1}}/4$). In steady-state the relationship between input and output frequencies of the divider of Fig. 6.3 is

$$f_{\text{in}} - \frac{1}{4} f_{\text{out1}} = 2 f_{\text{out1}} \tag{6.1}$$

and therefore

$$\frac{f_{\text{in}}}{f_{\text{out1}}} = 2.25. \tag{6.2}$$

In order to maximize hardware efficiency, the divide-by-2.25 and its subsequent divide-by-two are combined together. So, in Fig. 4.15, we propose a modified frequency divider architecture that achieves division ratios of both 2.25 and 4.5. As was discussed in Sect. 4.7, any frequency spur at the input of a divide-by-two translates into a spur at the output of the divider at the same offset frequency. So, the divider of Fig. 4.15 uses polyphase filtering to suppress the harmonics at the input of the feedback path dividers prior to mixing.

As was discussed earlier, an ILFD is chosen in the forward path of the divide-by-2.25 because it can operate a with smaller input drive, and also for its speed

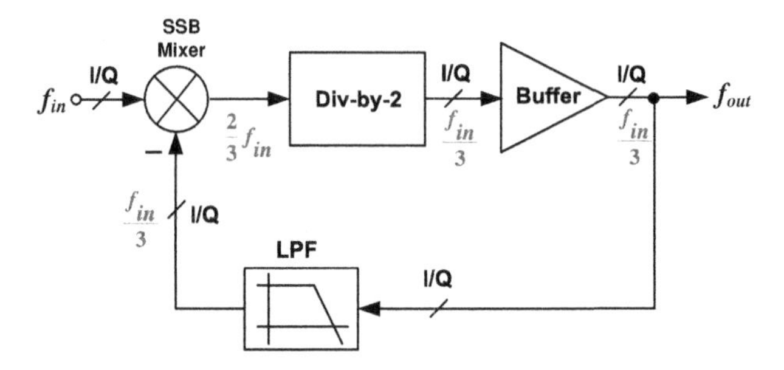

Fig. 6.4 Block diagram of the injection-locked regenerative divide-by-three with a 50 % duty cycle quadrature outputs

advantage over a static frequency divider. The stability and locking range of this frequency divider was discussed in Chap. 4.

The injection-locked regenerative divide-by-three of Fig. 6.4 is also designed using an architecture and design technique similar to that described in Sect. 4.7. This divide-by-three circuit is composed of a CML divide-by-two in the forward path and a low-pass filter in the feedback path. In spite of the use of conventional divide-by-three circuits, this divider can provide a 50 % duty cycle quadrature outputs.

The frequency divider of Fig. 6.4 receives a quadrature 1.584 GHz input signal, and generates a quadrature 528 MHz signal. The stability criteria and the locking range of this frequency divider can be obtained using similar techniques that are described in Chap. 4. The principles of operation of this frequency divider are similar to the frequency divider of Fig. 4.15. A single-sideband mixer is used to keep the difference of the input frequency, f_{in}, and the output of the feedback dividers, f_{out}. In steady-state, the relationship between input and output frequencies of this frequency divider is

$$f_{\text{in}} - f_{\text{out}} = 2f_{\text{out}} \tag{6.3}$$

and therefore

$$\frac{f_{\text{in}}}{f_{\text{out}}} = 3. \tag{6.4}$$

The second output of the divide-by-2.25 (f_{out2}) is fed to the divide-by-three of Fig. 6.4 to generate the channel spacing for operation of $SSB\ Mixer1$.

6.2.2 Linearity Considerations

An architecture based on the method of frequency division and mixing may suffer from spurious tones arising from the mixing. One method for spurious tone mitigation is a combination of single-sideband mixers with tuned loads [2]. However, in the absence of on-chip inductors, or generally any frequency selective load at the output of the mixers, the best approach is the use of single-sideband mixers along with

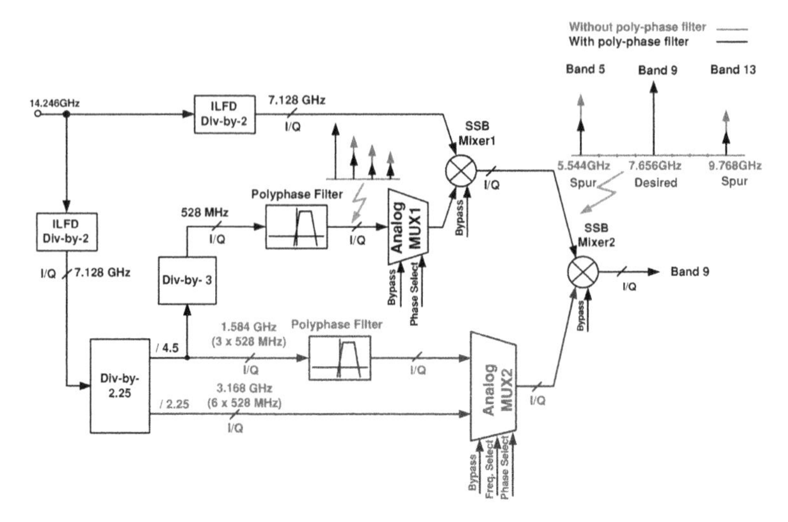

Fig. 6.5 The effect of polyphase filter on sideband reduction

linearization techniques. In this case, not only must the mixer be linearized, but the harmonic content of the inputs to the mixer must also be reduced [6].

As was discussed earlier, the outputs of the frequency dividers are used for frequency mixing. These signals are rich in harmonics. The worst case harmonic output is for the divide-by-three in Fig. 6.1, which is at 528 MHz. When the harmonics of 528 MHz contribute to mixing, they generate the center frequency of other bands. Depending on the process corner and the temperature, the kth harmonic of the divider output has a frequency roll-off proportional to $1/k^2$ (for a triangular waveform), or proportional to $1/k$ (for a square waveform). System simulations based on these waveforms reveal that at least 20 dB of attenuation of the third harmonic of the divide-by-three output is required in order to meet the spurious level requirement of MB-OFDM UWB. In this frequency synthesizer, the output of the divide-by-three and the output of the divide-by-4.5 are filtered prior to mixing. As will be discussed in Sect. 6.2.3, a combination of RC polyphase filters meets these requirements. Figure 6.5 shows a simulation of the spurious tones at the mixer output with and without polyphase filter spur reduction.

Mixer linearization is also a key in meeting the spurious specifications, and will be discussed in Sect. 6.2.4.

6.2.3 Polyphase Filter

As was discussed in Sect. 6.2.2, the filter preceded by the divide-by-three needs to provide over 20 dB suppression of the third harmonic of 528 MHz. To meet the emission requirements of MB-OFDM UWB, other harmonics of 528 MHz need to

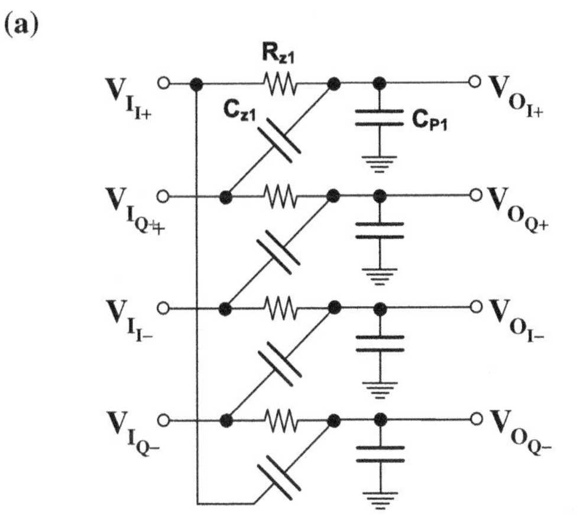

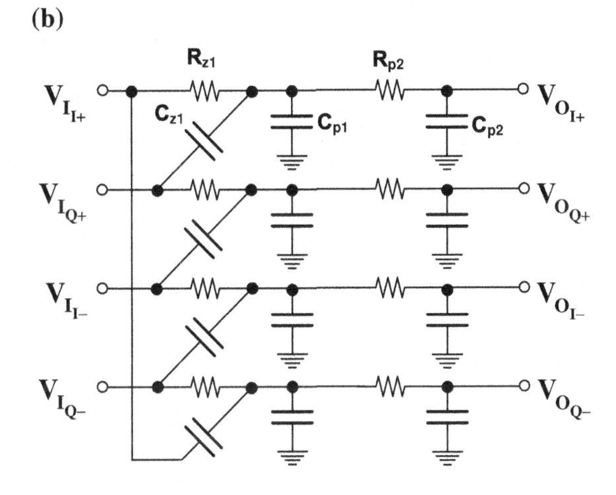

Fig. 6.6 **a** Implementation of an imaginary zero; **b** adding a real pole to further improve rejection

be lower than −40 dBc with respect to the fundamental. One way to do this is by generating a filter with imaginary zeros to cancel the harmonics. When on-chip inductors are not available, an alternative method is to use RC polyphase filters [7, 8]. This method is available, since quadrature phases at all the intermediate and output nodes are available.

Figure 6.6a shows an RC polyphase filter. The relationship between the inputs and outputs of this filter is

$$\begin{bmatrix} V_{O_{I+}}(s) \\ V_{O_{Q+}}(s) \\ V_{O_{I-}}(s) \\ V_{O_{Q-}}(s) \end{bmatrix} = \begin{bmatrix} a(s) & b(s) & 0 & 0 \\ 0 & a(s) & b(s) & 0 \\ 0 & 0 & a(s) & b(s) \\ b(s) & 0 & 0 & a(s) \end{bmatrix} \begin{bmatrix} V_{I_{I+}}(s) \\ V_{I_{Q+}}(s) \\ V_{I_{I-}}(s) \\ V_{I_{Q-}}(s) \end{bmatrix} \tag{6.5}$$

where $a(s)$ and $b(s)$ are defined as

$$a(s) = \frac{1}{1 + sR_{z1}(C_{z1} + C_{p1})} \tag{6.6a}$$

$$b(s) = \frac{sR_{z1}C_{z1}}{1 + sR_{z1}(C_{z1} + C_{p1})}. \tag{6.6b}$$

When the filter shown in Fig. 6.6a is driven by quadrature inputs, its transfer function can be simplified as

$$\frac{V_O}{V_I}(s) = \frac{1 \pm jsR_{z1}C_{z1}}{1 + sR_{z1}(C_{z1} + C_{p1})}. \tag{6.7}$$

The $\pm$ in (6.7) depends on the quadrature input sequence. The plus corresponds to a counterclockwise sequence of the inputs, while the minus sign corresponds to a clockwise sequence. The transfer function in (6.7) has an imaginary zero at $\pm j/R_{z1}C_{z1}$ and a real Left Half Plane (LHP) pole at $-1/R_{z1}(C_{z1} + C_{p1})$. At high frequencies, the gain of this transfer function is $C_{z1}/(C_{z1} + C_{p1})$. To further reduce the gain at high frequencies, another pole is added to the polyphase filter. This new filter is shown in Fig. 6.6b. The transfer function can be described by (6.5), where $a(s)$ and $b(s)$ are

$$a(s) = \frac{1}{a_2s^2 + a_1s + 1} \tag{6.8a}$$

$$b(s) = \frac{sR_{z1}C_{z1}}{a_2s^2 + a_1s + 1}. \tag{6.8b}$$

The a_1 and a_2 coefficients in (6.8a) and (6.8b) are

$$a_1 = R_{z1}(C_{z1} + C_{p1} + C_{p2}) + R_{p2}C_{p2} \tag{6.9a}$$

$$a_2 = R_{z1}R_{p2}(C_{z1} + C_{p1})C_{p2}. \tag{6.9b}$$

When quadrature inputs are applied to the input of Fig. 6.6b, its transfer function can be simplified to

$$\frac{V_O(s)}{V_I(s)} = \frac{1 \pm jsR_{z1}C_{z1}}{a_2s^2 + a_1s + 1} \tag{6.10}$$

When compared to Fig. 6.6a, the location of the imaginary zero is unchanged, and the filter achieves two LHP real poles. Moreover, the magnitude of the transfer function of the polyphase filter of Fig. 6.6b drops at high frequencies, which is desirable to suppress higher order harmonics. This filter is suitable for the output of the divide-by-4.5 because the output of this frequency divider (1.584 GHz) is triangular shaped. Therefore, the kth harmonic of the output has a frequency roll-off proportional to $1/k^2$ and the third harmonic is roughly 20 dB smaller than the fundamental, and the fifth harmonic is roughly 28 dB lower than the main harmonic. As a result, and from simulations, a single imaginary zero and two real poles provide sufficient attenuation of higher order harmonics.

In addition, the gain of the multiplexers and the Gm stage of the mixer is lower for the harmonics of 1.584 GHz. This relaxes the design requirements of the filter after the divide-by-4.5 circuit. On the other hand, these two conditions are not true for the output of the divide-by-three (528 MHz). Therefore, a higher order filtering is required for that divider.

Frequency dividers usually generate quadrature output phases by delaying the output by a quarter of the period. An example is the A D-latch-based divide-by-two of Fig. 6.7a. The relationship between the input and the outputs of this D-lattch-based divide-by-two is shown in Fig. 6.7a. Delaying a periodic signal by a quarter of its period leads to a phase shift of $-n\pi/2$ of its nth harmonic. A special case is a waveform with odd symmetry where only the odd harmonics are present. In this case the $(4k+1)$st harmonics of the frequency divider output have a counterclockwise quadrature sequence $\left(\exp(+j\pi/2)\right)$, while the $(4k+3)$rd harmonics have a clockwise quadrature sequence $\left(\exp(-j\pi/2)\right)$, where k takes any positive integer number including zero. This is illustrated in Fig. 6.7b. Therefore, when the harmonics are significant, we need at least two imaginary zeros: one for each of the quadrature sequences.

The complicated frequency dividers of Figs. 6.3 and 6.4 achieve quadrature output phases by using a divide-by-two at their outputs. Consequently, the harmonics of their output waveforms has a similar quadrature phase sequence as shown in Fig. 6.7b.

A higher order polyphase filter can be obtained by cascading the filters shown in Fig. 6.6a, b. If one inserts a buffer between these stages, the location of the poles and zeros of the overall transfer function will be the same as in the individual stages. However, this becomes challenging in a CMOS implementation since CMOS source-followers are quite lossy at these frequencies [9].

As shown in Appendix D, when cascading the polyphase filters of Fig. 6.6a, b—even without inserting a buffer between the stages—the location of the imaginary zeros is preserved. This eases the design and control of the zeros of the overall transfer function. However, the poles of the combined polyphase filter will change as a result of the interconnection.

On the other hand, cascading these passive RC stages increases the insertion loss of the overall filter. Here, two polyphase stages, similar to the one shown in Fig. 6.6a but with different quadrature sequences, are used to achieve adequate attenuation of all the close-in harmonics. In order to increase the roll-off of the mixer at high frequencies, an extra pole is added to the network, as shown in Fig. 6.6b. The resultant polyphase filter is shown in Fig. 6.8a.

If we assume that the input of Fig. 6.8a is driven by quadrature inputs, its transfer function can be written as

$$\frac{V_O}{V_I}(s) = \frac{(1 \pm jsR_{z1}C_{z1})(1 \mp jsR_{z2}C_{z2})}{a_3s^3 + a_2s^2 + a_1s + 1} \tag{6.11}$$

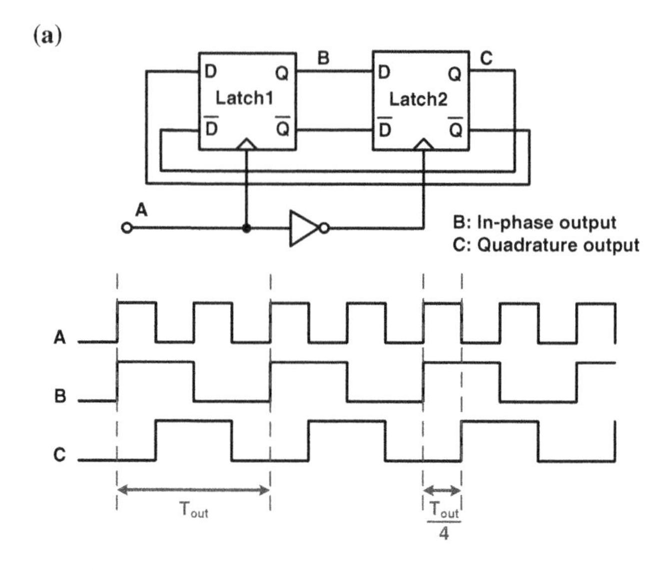

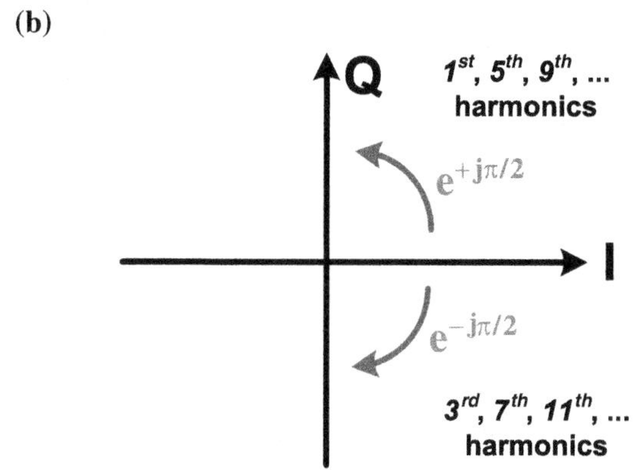

Fig. 6.7 **a** A D-latch based divide-by-two and its input and output waveforms; **b** I/Q phase sequence for different harmonics at the output of this divide-by-two

where the coefficients of the denominator of (6.11) are

$$a_1 = R_{p3}C_{p3} + R_{z1}(C_{z1} + 2C_{z2} + C_{p1} + C_{p2} + C_{p3}) + R_{z2}(C_{p2} + C_{z2} + C_{p3}) \tag{6.12a}$$

$$a_2 = R_{z1}R_{z2}((C_{p1} + C_{z1})(C_{p2} + C_{p3} + C_{z2}) + C_{z2} (c_{p2} + C_{p3})) + R_{z1}R_{p3}C_{p3}(C_{z1} + 2C_{z2} + C_{p1} + C_{p2}) + R_{z2}R_{p3}C_{p3}(C_{p2} + C_{z2}) \tag{6.12b}$$

$$a_3 = R_{z1}R_{z2}R_{p3}C_{p3}((C_{p1} + C_{z1})(C_{p2} + C_{z2}) + C_{p2}C_{z2}). \tag{6.12c}$$

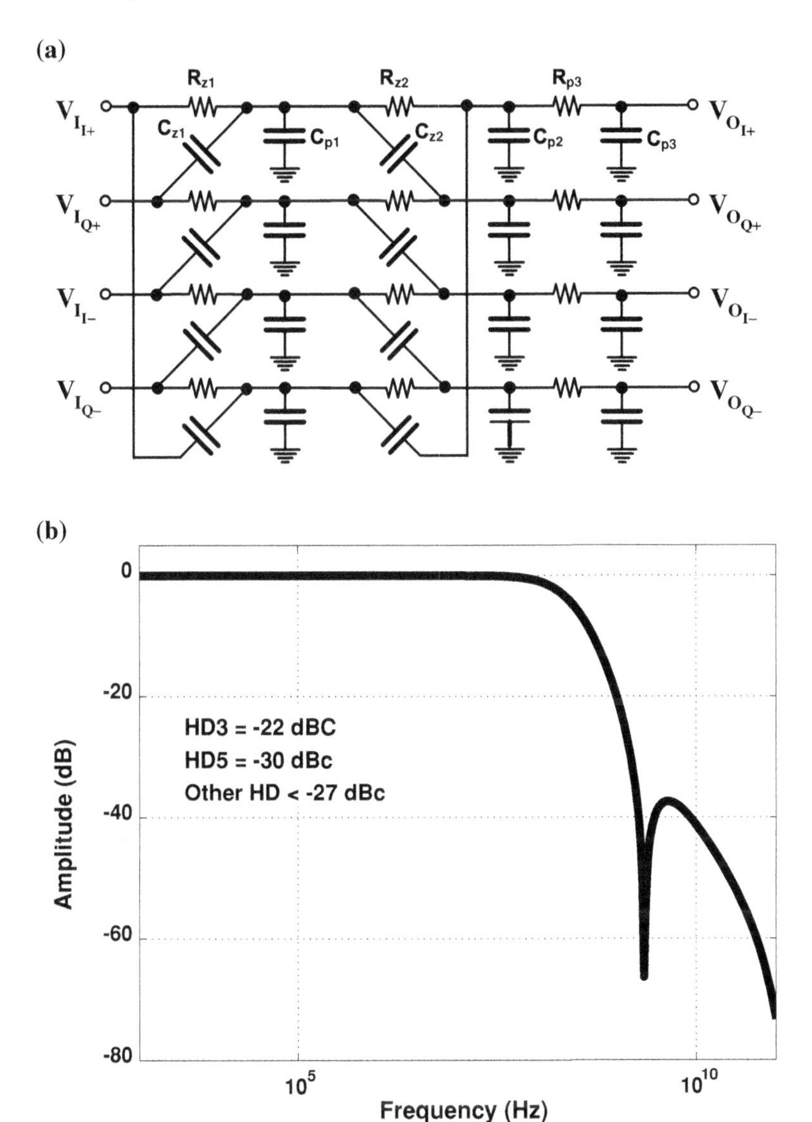

Fig. 6.8 Two-stage polyphase filter used to suppress the harmonics at the output of the divide-by-three. **a** Schematic; **b** simulated frequency response

The transfer function in (6.11) has imaginary zeros at $\pm j/R_{z1}C_{z1}$ and $\mp j/R_{z2}C_{z2}$. When $R_{z1} = R_{z1} = R_z$ and $C_{z1} = C_{z2} = C_z$, this filter has a pair of complex conjugate imaginary zeros at $\pm j/R_z C_z$. This guarantees a symmetric response for the clockwise and counterclockwise harmonics while leveraging the properties of the polyphase filters in implementing imaginary zeros. The frequency response of

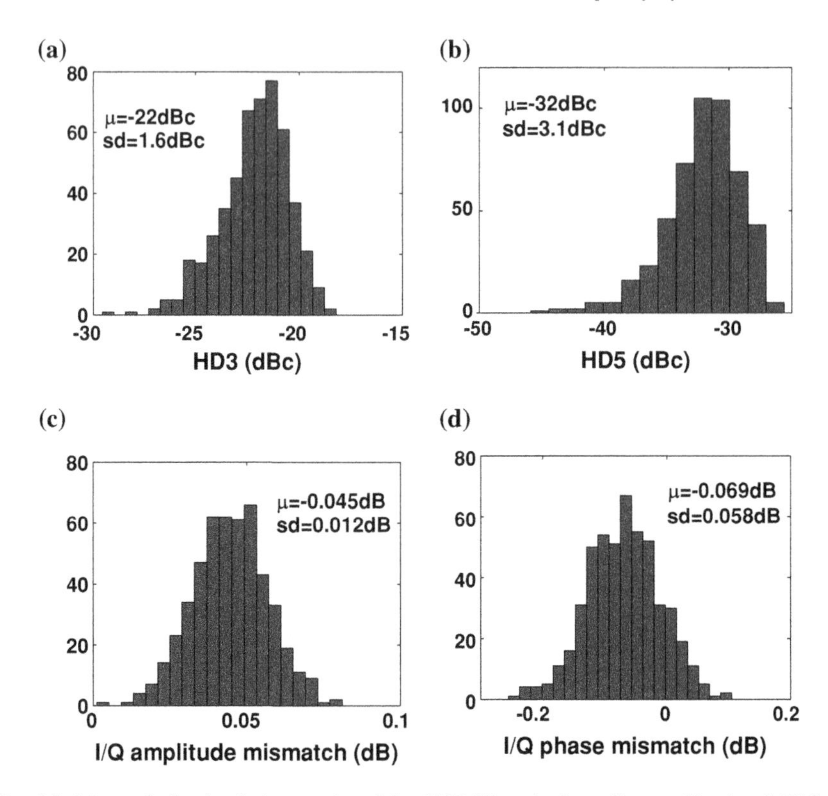

Fig. 6.9 Monte Carlo simulation results of the 528 MHz polyphase filter. **a** Simulated HD3; **b** simulated HD5; **c** simulated differential I/Q amplitude mismatch; **d** simulated differential I/Q phase mismatch

the polyphase filter of Fig. 6.8a is shown in Fig. 6.8b. This shows more than 22 dB attenuation for the third harmonic of 528 MHz, and roughly 30 dB attenuation of its fifth harmonic. All other harmonics are attenuated by more than 27 dB.

To make the polyphase filters less sensitive to the parasitics, all the capacitors are chosen to be larger than 200 fF. Monte Carlo simulations are performed to examine the effects of process variation and mismatch on the performance. Figure 6.9 shows the simulated results of 500 runs of Monte Carlo simulation of the polyphase filter of Fig. 6.8a. As can be seen, the mean value of the HD3 is −22 dBc with a standard variation of 1.6 dBc. It is also observed that the I/Q amplitude and phase mismatches at the output of the polyphase filter respectively remain less than 0.1 dB and 0.2°. Similar Monte Carlo simulations are performed for the polyphase filter of Fig. 6.6b. In this case the mean value of HD3 is −22 dBc with a standard variation of 3.1 dBc, while the mean value of HD5 is −16 dBc with standard variation of 0.9 dBc. The I/Q amplitude and phase mismatches at the output of this polyphase filter are respectively less than 0.03 dB and 0.5°. So, the process variation will not significantly degrade the performance of the polyphase filters.

6.2.4 Single-Sideband Mixer

The synthesizer uses differential quadrature signals throughout, so that single-sideband mixers can be employed. The function of the single-sideband mixers is to implement in-phase and quadrature phases of the sum or difference of two frequencies. $SSB\ Mixer1$ generates the center frequencies for the bands in Group 3. Depending on the target band and group, $SSB\ Mixer2$ will down-convert, up-convert or leave unchanged the output of $SSB\ Mixer1$.

Most of the frequency synthesizers that are implemented based on this method use inductive loads in the single-sideband mixers [2, 10, 11]. This is desirable for an efficient use of the headroom, achieving a large signal swing, and suppression of the spurious responses. However, when a resistive load is used at the output of the mixers, especially at 1.2 V or lower, the headroom problems becomes more challenging. To overcome headroom problems, the Gm stage of $SSB\ Mixer1$ is folded (Fig. 6.10), mirrored, and fed to the NMOS switching quad. In this case, the signal swing at the output of the mixer will not cause the Gm stage to enter the triode region. Moreover, the output stage has a greater headroom, and achieving a larger signal swing at the output is feasible. But it increases the power consumption.

Both single-sideband mixers can be bypassed if necessary (for example, for synthesis of band 8), by reconfiguring them as a differential pair that passes the LO frequency. Transistors $Ma1$ and $Ma2$ in Fig. 6.10 sink dc current to maintain a constant common-mode DC level in the two modes. This technique, along with a fast settle DC bias circuit that turns on $Ma1$ and $Ma2$ very quickly, results in a switching time of less than 2 nS for most of the switching scenarios. To adjust for I/Q mismatch, quadrature correction is embedded inside the LO buffer prior to $SSB\ Mixer1$ and $SSB\ Mixer2$. One remaining spurious output is the (fixed) LO to output leakage, which occurs through layout asymmetry, systematic offset in the mixers, and inductive coupling in the power supply lines, and which is roughly −20 dBc at 7.128 GHz. This can be further reduced through improved high-frequency packaging.

Because of its stringent linearity requirements $Analog\ Mux1$ is implemented using resistively degenerated differential pairs. $Analog\ Mux2$ selects the correct phase as well as the appropriate frequency. As shown in Fig. 4, it is composed of four Gm stages. In order to achieve sufficient swing with the low power supply, and also because of its moderate linearity requirements, inverter-based Gm stages are used for $Analog\ Mux2$. When $SSB\ Mixer2$ is in bypass mode, the Gm stages in $Analog\ Mux2$ are switched off and the output of $Analog\ Mux2$ is shorted to ground to provide better isolation.

The input frequencies to the $SSB\ Mixer1$ are 7.128 GHz and 528 MHz. Reduction of the spurious tones, due to 528 MHz signal, at the output of this mixer implies that: (1) the 528 MHz signal must have low harmonic content, (2) a linear Gm stage is needed in the mixer to avoid generating harmonics, and (3) a small amount of I/Q amplitude and phase mismatch is caused. The first requirement is met by means of polyphase filtering of the harmonics, as was discussed in Sect. 6.2.3. The I/Q amplitude and phase matching of (3) is achieved by careful symmetric layout.

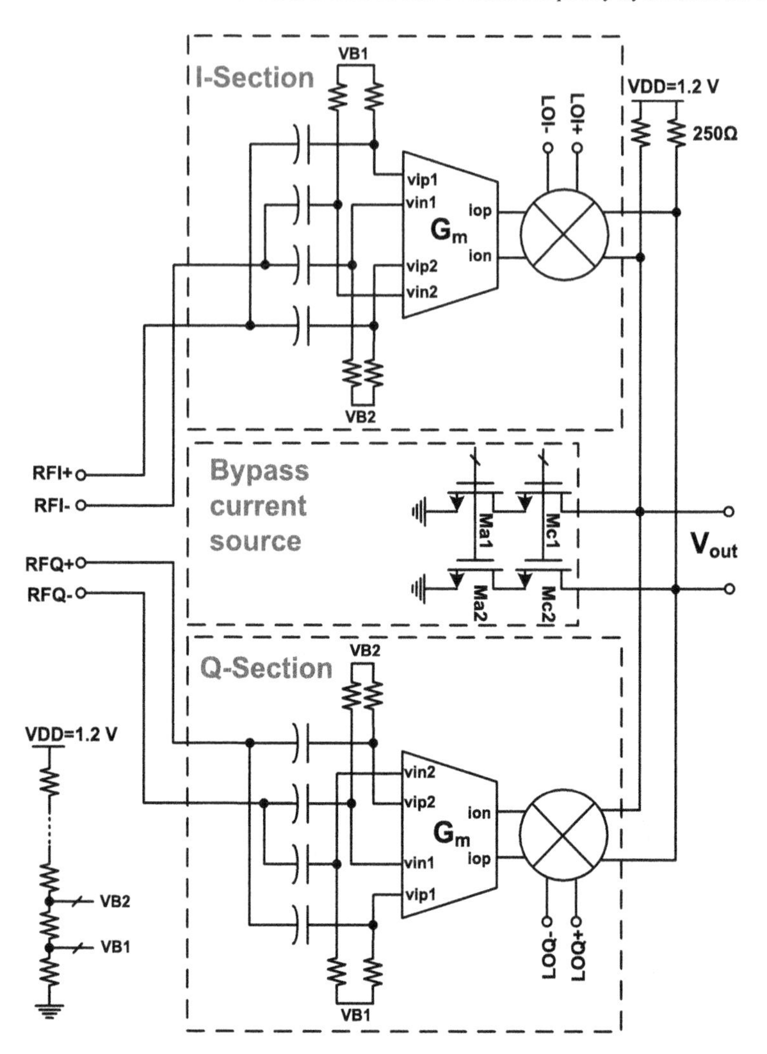

Fig. 6.10 Single-sideband mixing stage

The second requirement is achieved by linearizing the Gm stage of the mixer. Among all different methods of Gm linearization, we require a method that is suitable for low supply voltage and high frequencies, and yet does not reduce the conversion gain. Source-degeneration [12] leads to a reduction of the conversion gain of the mixer. In this work, the Gm stage of the mixer of Fig. 6.10 is linearized by using a multi-tanh Gm stage [13]. This technique provides a linear input range and does not degrade the gain of the Gm stage. The Gm stage is shown in Fig. 6.11a, and the overall linearized transconductance is plotted in Fig. 6.11b. Voltages $VB1$ and $VB2$ in Fig. 6.10 are chosen to maximize the gain flatness for the desired input voltage range.

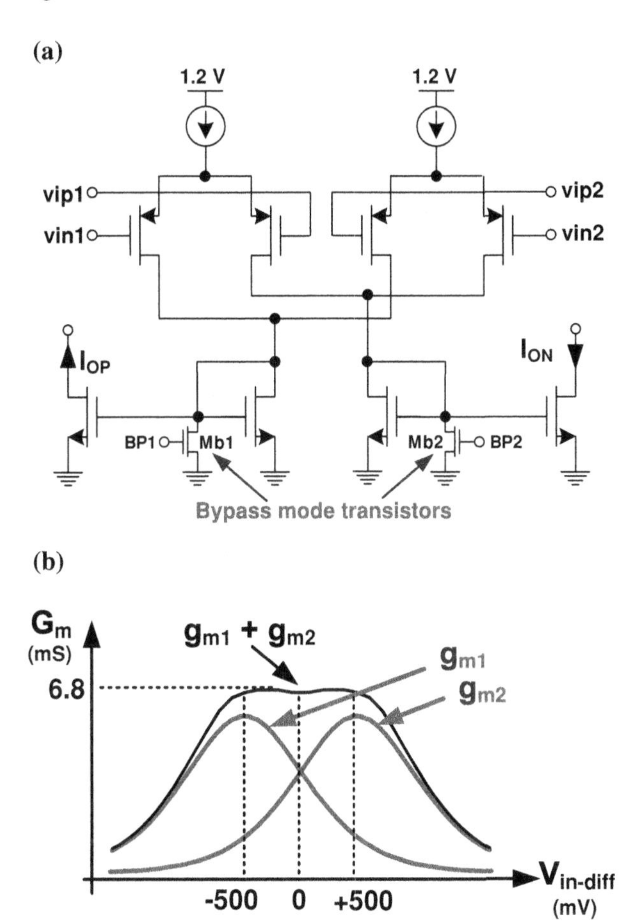

Fig. 6.11 Mixer linearization using multi-tanh Gm stage. **a** Multi-tanh Gm stage with improved headroom; **b** linearized Gm

A fast bias circuit shown in Fig. 6.12b is used to reduce the time it takes the mixers to go to the bypass mode. Compared to the conventional bias circuit of Fig. 6.12a, transistors $M4$, $M5$, and $M6$, are added. Transistor $M5$ in the fast-settle bias circuit never enters the triode region, which speeds its switching. When the bypass goes low, $M5$ charges up the gate of $M1$, node A, by injecting extra current. When the voltage at node A reaches the desired value, $M5$ turns off.

Figure 6.12c shows the simulation result of these two bias circuits. The fast settle bias circuit of Fig. 6.12b enables this frequency synthesizer to achieve a switching time of less than 2 nS in all the band switching scenarios.

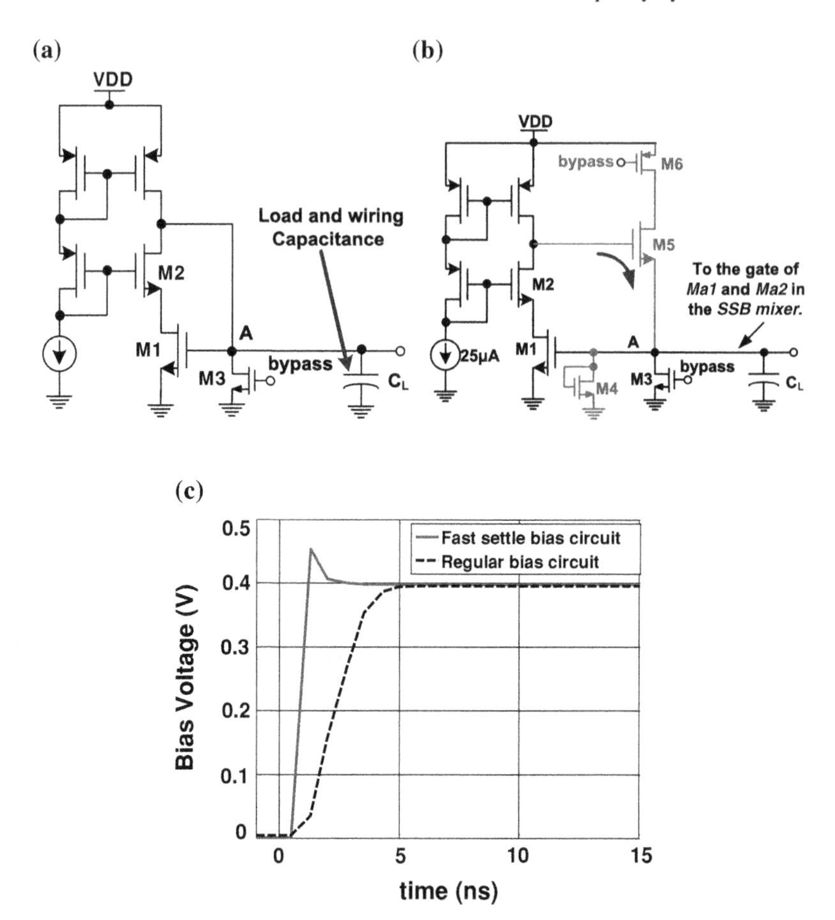

Fig. 6.12 **a** Regular bias circuit; **b** fast-settle bias circuit; **c** comparison of startup time of regular bias circuit and fast bias circuit

6.2.5 Mixer Interface

As was discussed in Sect. 6.2.4, this synthesizer uses up to two levels of single-sideband mixing. In addition, based on the comparative study of different schemes for cascading two mixers that was presented in Sect. 5.2.1 it can be concluded that the voltage-mode solution is the preferred solution for a CMOS inductor-less implementation of UWB synthesizer with a supply voltage of 1.2 V or lower. Consequently, this implementation is used here. As can be seen from Fig. 6.13, the output of the first single-sideband mixer goes to the LO port (switches) of the second single-sideband mixer. This method of interfacing has two main advantages: (1) it allows linearization of the Gm stage of $SSB\ Mixer2$, which is vital to meet the out-of-band emission requirements. (2) Since all the odd harmonics of the LO signal contribute

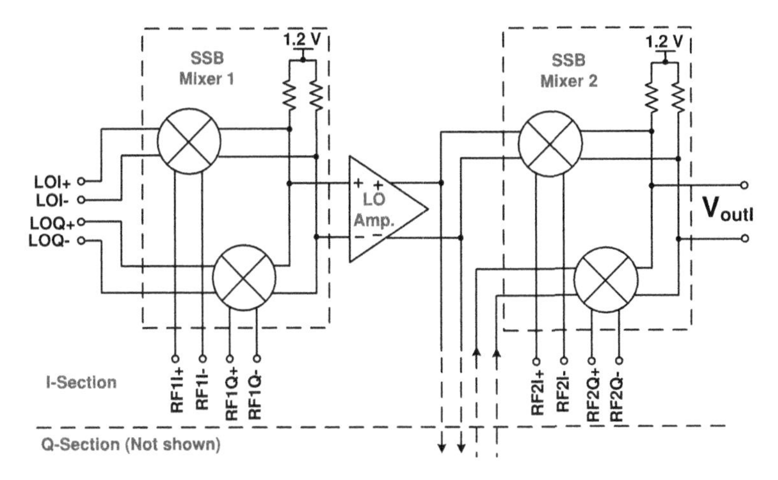

Fig. 6.13 Mixing stages and interface amplifier

to the mixing products, the higher frequency signal is applied to the LO port of $SSB\ Mixer2$. In the proposed frequency plan, this signal can be at 6.6, 7.128, or 7.656 GHz,
depending on the synthesized band. As a result, higher order mixing products, due to higher order harmonics of LO signal, lie outside the UWB span, and at the same time, experience low gain at those frequencies.

This scheme also eases the design and linearization of multiplexers used for band switching. One interfacing problem is the large gain-bandwidth product requirement of the interface amplifier shown in Fig. 6.13. The large gain is required so it does not degrade the conversion gain, as well as the noise performance. In this work we use our proposed inductor-less CMOS broadband amplifier that was introduced in Sect. 5.4. The amplifier is shown in Fig. 5.14a and consists of three gain stages and is followed by an output buffer. The schematic of the gain stage used in this amplifier is shown in Fig. 5.14b and the schematic of the output buffer is depicted in Fig. 5.14a. The overall frequency response of this composite amplifier is shown in Fig. 5.15.

6.2.6 Multiplexer Design

A fully quadrature implementation of the frequency synthesizer requires two of the single-sideband mixers shown in Fig. 6.10 in each mixing stage. If we refer to these mixers as $SSB\ Mixer_I$ and $SSB\ Mixer_Q$, and also assume that the signals applied to the LO port and Gm stage of the single-sideband mixers are respectively at the frequency of ω_{LO} and ω_{RF}, then the following sequence shown in Table 6.1 needs to be applied to the mixer in order to achieve the $\cos((\omega_{LO} \pm \omega_{RF})t)$ at the output of $SSB\ Mixer_I$ and $\sin((\omega_{LO} \pm \omega_{RF})t)$ at the output of $SSB\ Mixer_Q$. In Table 6.1,

Table 6.1 Operation of multiplexers

	SSB mixer$_I$	SSB mixer$_Q$
LOI	$\cos(\omega_{LO}t)$	$\sin(\omega_{LO}t)$
LOQ	$\sin(\omega_{LO}t)$	$\cos(\omega_{LO}t)$
RFI	$\cos(\omega_{RF}t)$	$\cos(\omega_{RF}t)$
RFQ	$\mp\sin(\omega_{RF}t)$	$\pm\cos(\omega_{RF}t)$

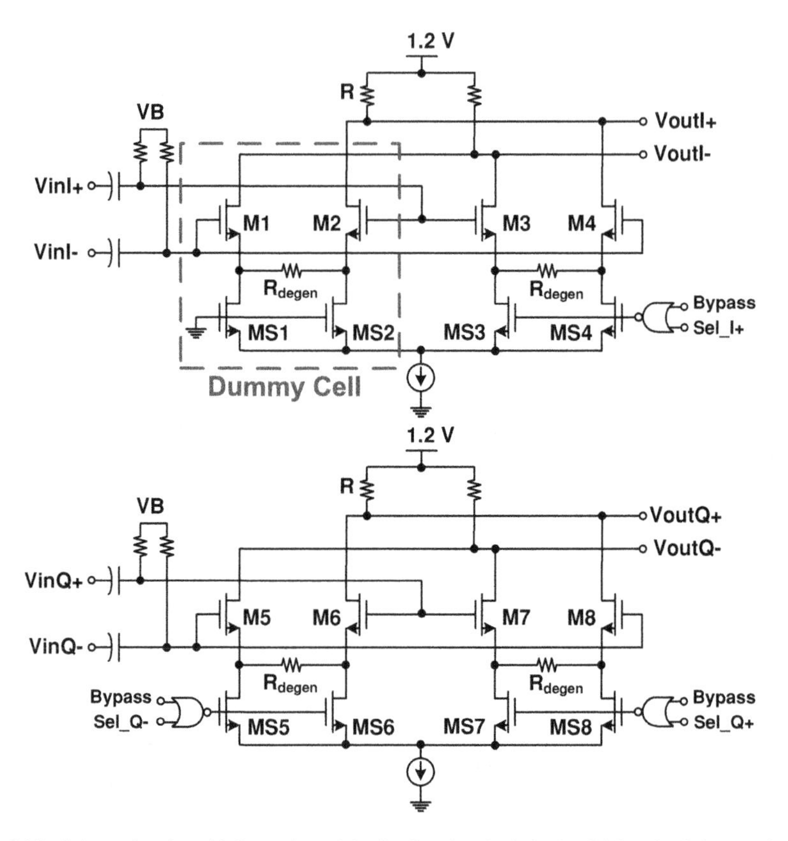

Fig. 6.14 Schematic of multiplexer 1 used in the first level of single-sideband mixing to choose the appropriate phase of the 528 MHz signal

LOI and LOQ are the differential signals applied to the LO port of the single-sideband mixer shown in Fig. 6.10, while RFI and RFQ are the differential signals applied to its Gm stage.

As can be seen in Table 6.1, in order to perform the band switching, only the signal to the Gm stage of the mixer undergoes a change in its polarity, and the LO signal is always fixed. In Table 6.1, $\mp$ for $\sin(\omega_{RF}t)$ and $\pm$ for $\cos(\omega_{RF}t)$ are chosen to generate the sum or the difference of two frequencies, respectively. This method is used in both multiplexers for both stages of mixing. This method eases the implementation of the multiplexers, since the signal applied to the Gm stage of each mixer

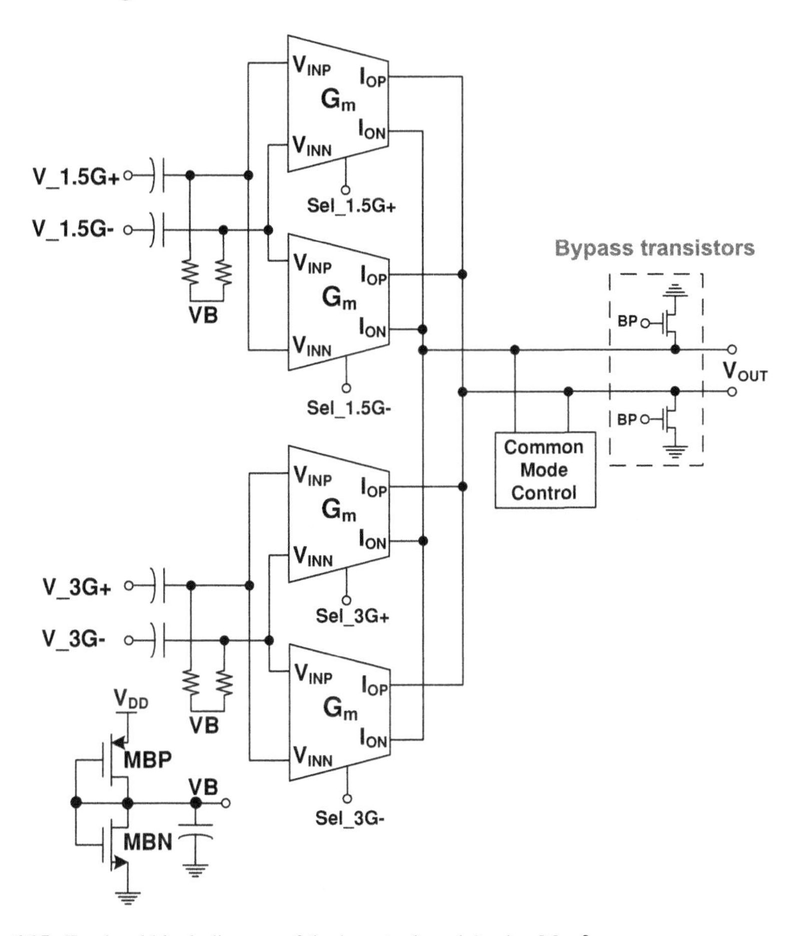

Fig. 6.15 Top level block diagram of the inverter-based Analog Mux2

is always at a lower frequency than the LO signal. On the other hand, in the second stage of mixing, *Analog Mux*2 must be able to apply different frequencies to the Gm stage. Therefore, this method combines the phase and frequency selection in one block. *Analog Mux*1 accomplishes this function for the first level of mixing. A quadrature differential 528 MHz is applied to the *Analog Mux*1, and it applies the appropriate phases of 528 MHz to the Gm stage of the single-sideband mixers in the first level of mixing. Due to its high linearity requirement, *Analog Mux*1 is linearized using a source-degenerated structure [12]. The simplified schematic of *Analog Mux*1 is shown in Fig. 6.14.

*Analog Mux*2 has to provide the appropriate phase sequences of the signal, but it also has to provide the right frequency for the operation of *SSB Mixer*2. In order to achieve a compact implementation, and also to achieve large gain, inverter-based Gm stages [14] are used (Figs. 6.15, 6.16).

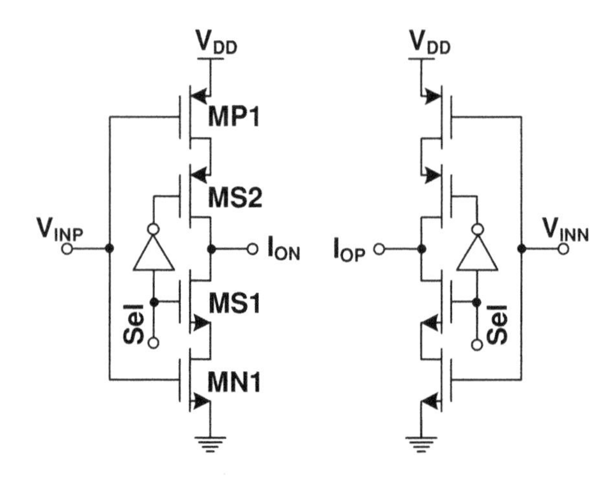

Fig. 6.16 Pseudo-differential inverter-based Gm stage used in Analog Mux2

6.3 Measured Results

Figure 6.17 shows the measured band switching of the frequency synthesizer. In contrast to many of the reported UWB frequency synthesizers, this design is capable of hopping from any arbitrary UWB band to any other. It also supports operation in each of the individual UWB Band Groups (1–6) shown in Fig. 1.9. Two different band switching scenarios are shown in Fig. 6.17: switching within bands of a group, and switching between bands of different groups. The synthesizer settles in approximately 2 nS when hopping from band 1 to band 2, and settles in approximately 1 nS when hopping from band 11 to band 5. Further measurement results show that this design achieved settling times of less than 2 nS when hopping between any of the other UWB bands.

Figure 6.18a, b show the measured output spectrum of frequency band #7 within Band Group three and in the entire UWB span, respectively. Note that the increase in the noise floor in the 3.1–6.0 GHz band is due to the measurement setup. The measurement results of other bands have similar spectra.

The in-band spurious tones for all 14 bands are shown in Table 6.2. As can be seen from this table, the mixing sidebands are better than −30 dBc. In addition, as mentioned earlier, there is a fixed LO to output leakage at 7.128 GHz which is smaller than −20 dBc, depending on the Band Group. This spurious tone occurs through inductive coupling in the power supply lines, layout asymmetry in single-sideband Mixer1, and systematic offset in the mixers.

To clarify the coupling in the power supply and bias lines, the photograph of the packaged chip, using a 7 × 7 mm 48-lead MLF-QFN package, is shown in Fig. 6.19. As can be seen from this Fig. 6.19, the package body size is significantly larger than the chip dimensions. As a result, the bonding wires are very long and major crosstalk occurs among the supply lines that leads to an unwanted spurious tone at 7.128 GHz.

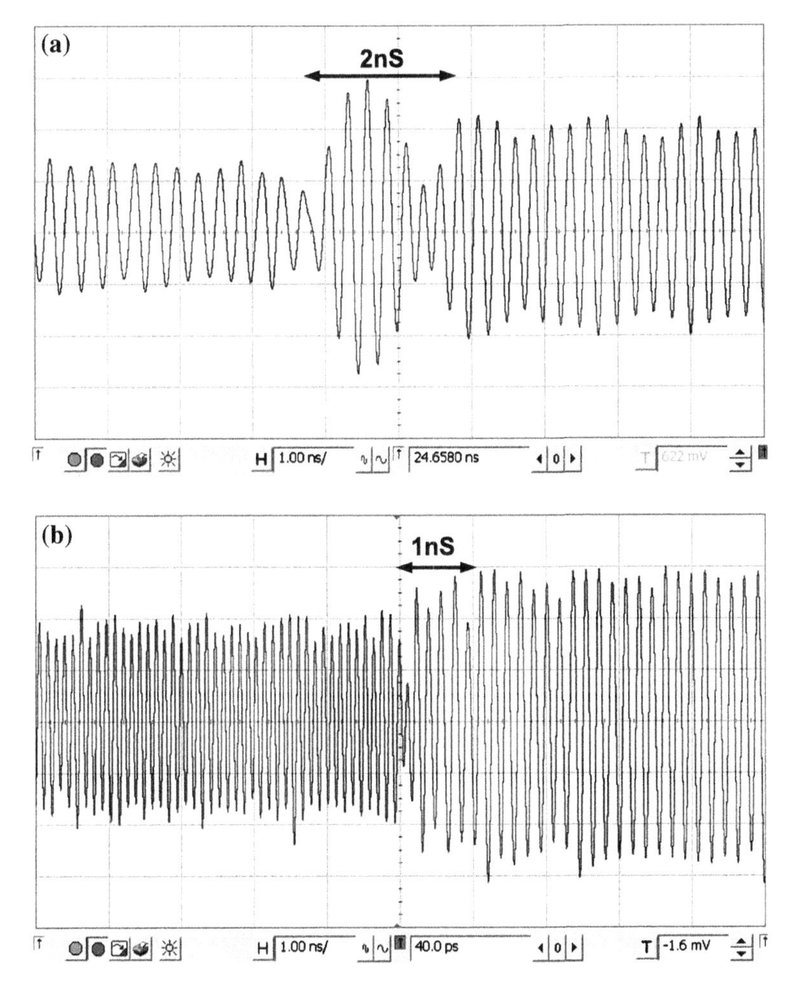

Fig. 6.17 Measured band switching time. **a** band switching within a group (band #1 to band #2); **b** band switching between groups (band #11 to band #5)

This can be further reduced through improved high-frequency packaging or using chip on board assembly to shrink the length of the bonding wires.

Another spurious tone that can be seen in Table 6.2 is the LO to output leakage in the second single-sideband mixer which is roughly −28 dBc or better, depending on the band and Band Group of operation. This spurious tone also occurs due inductive coupling in the power supply lines, layout asymmetry in single-sideband Mixer2, and systematic offset in the mixers, and it could be reduced by improving the layout and using a high-frequency packaging technique.

Figure 6.20 shows the measured phase noise for different bands. The measured phase noise at 1 MHz offset is always better than −114 dBc/Hz. The phase noise of band 8 is the best of all, since generating this band requires no mixing. The phase noise

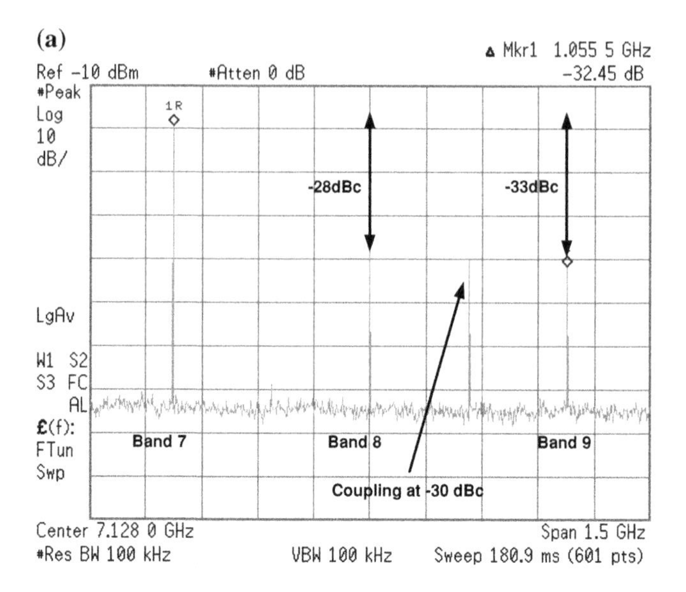

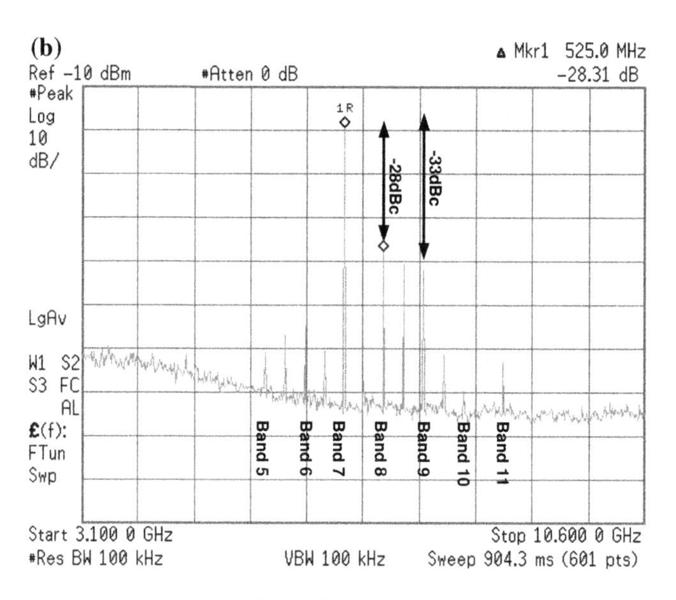

Fig. 6.18 Measured output spectrum of band #7 **a** spurious tones in group 3, and **b** spurious tones in the whole UWB span

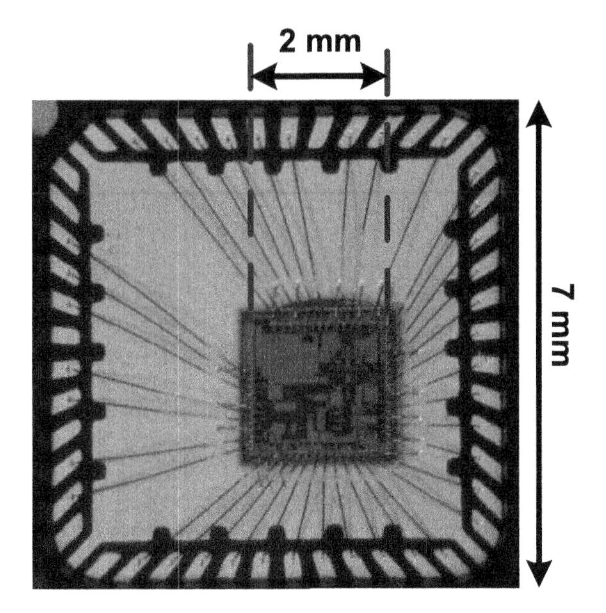

Fig. 6.19 The packaged chip using a 7 × 7 mm 48-lead MLF-QFN package

worsens as the number of mixing and division stages in the signal path increases. As a result, the phase noise of the center band in each group (bands 2, 5, 8, 11, and 14) is better than the phase noise of the other bands in that group.

Table 6.3 summarizes the measured results of the UWB synthesizer, and compares it with other published 14-band synthesizers. Compared to similar designs, the proposed synthesizer is implemented using minimal hardware, no on-chip inductors, and uses the lowest supply voltage. The inductor-less design methodology will lead to further area reduction when the design is migrated to a finer technology node.

Figure 6.21 shows the chip microphotograph. The core area is 1.3 mm^2. The chip is packaged using a 48 pin MLF-QFN package.

The frequency synthesizer of Fig. 6.1 requires a 14.256 GHz signal in order to generate all the required frequencies for the operation of its frequency dividers and mixers. Another variant of the frequency synthesizer of Fig. 6.1 is implemented where the 14.256 GHz signal is generated using an on-chip phase-locked loop (PLL). The block diagram of this variant is shown in Fig. 6.22.

In the architecture of Fig. 6.22 the PLL shares its prescalers (the ILFD divide-by-two, the divide-by-2.25, and the divide-by-three) with the ones required to implement the core frequency synthesizer of Fig. 6.1. This is done to reduce the die area and the power consumption. One of the problems of this prescalers sharing is the potential coupling from the multiplexers to the VCO. Any band switching requires change of the phase or frequency in the multiplexers, and in case of insufficient isolation the kick back can unlock the PLL. Consequently, buffers are inserted before and after the frequency dividers to increase the isolation to the VCO. The frequency synthesizer

Table 6.2 Spurious tones measured results

Target Band Group/band		Spurious tones		
		Band 1	Band 2	Band 3
	Band 1	–	−34 dBc	−34 dBc
Band Group 1	Band 2	<−60 dBc	–	<−60 dBc
	Band 3	−34 dBc	−28 dBc	–
		Band 4	Band 5	Band 6
	Band 4	–	−32 dBc	−39 dBc
Band Group 2	Band 5	<−55 dBc	–	<−55 dBc
	Band 6	−37 dBc	−28 dBc	–
		Band 7	Band 8	Band 9
	Band 7	–	−28 dBc	−33 dBc
Band Group 3	Band 8	No spurious tone	–	No spurious tone
	Band 9	−37 dBc	−23 dBc	–
		Band 10	Band 11	Band 12
	Band 10	–	−29 dBc	−31 dBc
Band Group 4	Band 11	<−55 dBc	–	<−55 dBc
	Band 12	−30 dBc	−30 dBc	–
		Band 13	Band 14	-
Band Group 5	Band 13	–	−37 dBc	–
	Band 14	<−55 dBc	–	–

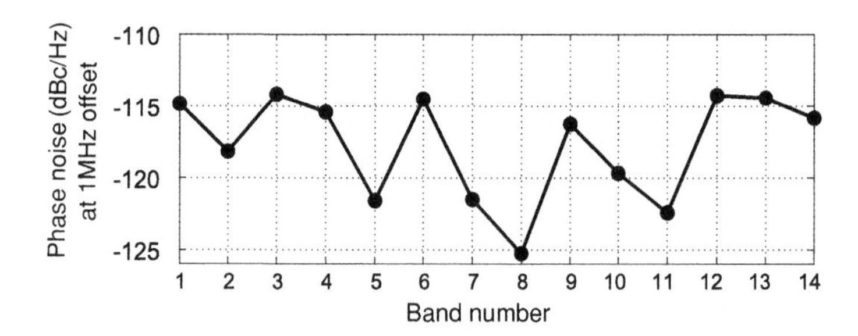

Fig. 6.20 Phase noise for different bands at 1 MHz offset

of Fig. 6.22 is implemented in a 0.13 μm CMOS process, and it achieves a similar switching time as the architecture of Fig. 6.1.

6.4 Conclusion

This work presented the first CMOS inductor-less single PLL 14-band frequency synthesizer for MB-OFDM UWB which is capable of performing any arbitrary band switching, or operate in any individual Band Groups shown in Fig. 1.9.

Table 6.3 Table of comparison with other 14-band frequency synthesizers

Reference	Liang [2]	Werther [11]	Lu [10]	This work
Technology (μm)	0.18 CMOS	0.13 SiGe	0.18 CMOS	0.13 CMOS
Required PLL's	2	1	1	1
On-chip inductors[a]	8	Not reported	8	0
SSB mixing levels	3	2	3	2
Switching time (nS)	<3	<3	<3	<2
Supply voltage (V)	1.8	2.4/1.2	1.8	1.2
Power consumption	160 mW	Not reported	117 mW	135 mW
Die area	1.5 mm^2[b]	Not reported	5.5 mm^2	1.3 mm^2[c]

[a]Minimum required number of on-chip inductors in the synthesizer part. The inductors in the VCO, and VCO buffer(s) are not taken into account
[b]Core area reported
[c]Core area reported

Fig. 6.21 Chip microphotograph

This synthesizer exploits up to two levels of single-sideband mixing and uses an external 14.256 GHz signal to generate all the required frequencies. It is implemented in a 0.13 μm CMOS process, uses a single 1.2 V supply voltage, and dissipates 135 mW. The mixing sideband level is better than −31 dBc. The synthesizer can perform frequency switching among all different bands and groups, and the switching time is roughly 2 nS for all different hopping scenarios. The phase noise is better than −110 dBc/Hz at 1 MHz offset.

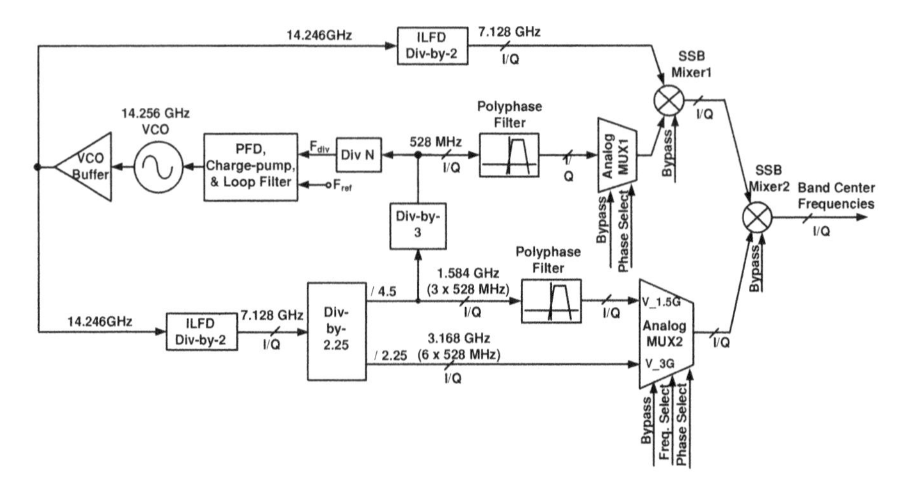

Fig. 6.22 Another proposed 14-band UWB frequency synthesizer with a PLL

References

1. Batra A, Balakrishnan J, Dabakand A et al (2004) Multi-band OFDM physical layer proposal for IEEE 802.15 Task Group 3a, March 2004
2. Liang C-F, Liu S-I, Chen Y-H, Yang T-Y, Ma G-K (2006) A 14-band frequency synthesizer for MB-OFDM UWB application. In: IEEE ISSCC digest of technical papers, Feb 2006, pp 428–437
3. Chominski P, Malhi D, Larson L, Park J, Gudem P, Kloczkowski R, Demirdag C, Garlapati A, Pereira V (2002) A highly integrated Si/SiGe BiCMOS upconverter RFIC For 3G WCDMA handset applications. In: Proceedings of the European solid-state circuits conference, Sept 2002, pp 447–450
4. Miller R (1939) Fractional-frequency generators utilizing regenerative modulation. Proc IRE 27(7):446–457
5. Lin C, Wang C-K (2005) A regenerative semi-dynamic frequency divider for mode-1 MB-OFDM UWB hopping carrier generation. In: IEEE ISSCC digest of technical papers, Feb 2005, vol 1, pp 206–207
6. Razavi B, Aytur T, Lam C, Yang F, Li K, Yan R, Kang H, Hsu C, Lee C (2005) A UWB CMOS transceiver. IEEE J Solid-State Circuits 40(12):2555–2562
7. Behbahani F, Kishigami Y, Leete J, Abidi A (2001) CMOS mixers and polyphase filters for large image rejection. IEEE J Solid-State Circuits 36(6):873–887
8. Galal S, Ragaie H, Tawfik M (2000) RC sequence asymmetric polyphase networks for RF integrated transceivers. IEEE Trans Circuits Syst II: Analog Dig Signal Process 47(1):18–27
9. Razavi B (2001) Design of analog CMOS integrated circuits. McGraw-Hill, Boston
10. Lu T-Y, Chen W-Z (2008) A 3-to-10 GHz 14-band CMOS frequency synthesizer with spurs reduction for MB-OFDM UWB system. In: IEEE ISSCC digest of technical papers, Feb 2008, pp 126–601
11. Werther O, Cavin M, Schneider A, Renninger R, Liang B, Bu L, Jin Y, Marcincavage J (2008) A fully integrated 14-band 3.1-to-10.6 GHz 0.13μm SiGe BiCMOS UWB RF transceiver. In: IEEE ISSCC digest of technical papers, Feb 2008, pp 122–601

12. Sanchez-Sinencio E, Silva-Martinez J (2000) CMOS transconductance amplifiers, architectures and active filters: a tutorial. IEEE Proc Circuits Devices Syst 147(1):3–12
13. Gilbert B (1998) Themulti-tanh principle: a tutorial overview. IEEE J Solid-State Circuits 33(1):2–17
14. Nauta B (1992) A CMOS transconductance-C filter technique for very high frequencies. IEEE J Solid-State Circuits 27(2):142–153

Chapter 7
Conclusion

The wide bandwidth and stringent agility requirements of MB-OFDM UWB poses several design challenges to the implementation of a monolithic UWB transceiver. Since the spectrum from 3.1 to 10.6 GHz is not universally available, any UWB solution needs to be able to operate in different Band Groups to be marketable globally. One major challenge in the implementation of any UWB transceiver that covers several Band Groups is to generate all the band center frequencies and meeting the agility requirements of WiMedia UWB as well as the spurious tone limitations imposed by FCC. On the other hand, UWB is targeting applications such as wireless USB that imply a very low-cost UWB solution. This can only be achieved when a hardware efficient architecture for a UWB radio is implemented in a standard CMOS technology and only using the standard features of a digital CMOS process. As a result, the focus of this book has been on

- Presenting an architecture for monolithic frequency synthesizer for MB-OFDM UWB that can operate in all six Band Groups of WiMedia UWB.
- Implementing the proposed architecture in a digital submicron CMOS technology without using any on-chip inductor or balun.

To overcome the settling time limitations of conventional frequency synthesizers and covering the entire Band Groups of UWB (a bandwidth of 7.5 GHz with 528 MHz channel spacing), an architecture based on the method of frequency division and mixing is used. The main challenge in using this architecture is to reduce the number of required PLLs and SSB mixers.

Since operation of SSB mixers requires quadrature signals, techniques for quadrature generation at microwave frequencies are investigated, and four-stage ring-oscillator-based injection-locked frequency dividers are studied as a technique to implement a frequency divider and achieve quadrature phases of output. Toward this goal, a multi-phase multi-modulus ring-oscillator-based ILFD is designed in a 0.13 μm CMOS technology that works as a divide-by-six at V-band, and can also achieve division ratios of four and two.

A frequency plan was proposed that can generate all the required frequencies from a single fixed frequency and can implement any center frequency with a maximum

M. Farazian et al., *Fast Hopping Frequency Generation in Digital CMOS*,

DOI: 10.1007/978-1-4614-0490-3_7,

of two levels of SSB mixing. In order to generate all the required frequencies for the operation of this frequency synthesizer out of a single frequency, fractional frequency dividers are needed. Therefore, another study was performed on the architectures that can obtain a fractional division ratio. This study involved an analysis of operation and stability of injection-locked regenerative frequency dividers. This was followed by a phase noise analysis of this class of frequency dividers.

In addition, the operation, stability, locking range, and phase noise of two-stage ring-oscillators, which are compact ways to generate quadrature output phases and can be used in injection-locked regenerative frequency dividers, was analyzed.

In order to meet the linearity requirement in an inductor-less design, low-voltage linearization techniques along with polyphase filtering are employed. To use polyphase filters for spurious tones mitigation, the behavior of polyphase filters in the presence of process variations is carefully examined.

Finally, this book presented a CMOS inductor-less single PLL 14-band frequency synthesizer for MB-OFDM UWB, which is capable to perform any arbitrary band switching, or operate in any individual Band Group shown in Fig. 1.9.

This synthesizer exploits up to two levels of SSB mixing and uses an external 14.256 GHz signal to generate all the required frequencies. It is implemented in a 0.13 μm CMOS process, uses a single 1.2 V supply voltage, and dissipates 135 mW. The mixing sideband level is better than -31 dBc. The synthesizer can perform frequency switching among all different bands and groups, and the switching time is roughly 2 nS for all different hopping scenarios. The phase noise is better than -110 dBc/Hz at 1 MHz offset.

This work presented the possibility of fast hopping high-speed signal generation in a digital CMOS technology. This methodology fully benefits from the technology scaling, and our proposed inductor-less design methodology leads to a smaller die area and lower power consumption when this design is ported to a finer technology node.

Appendix A
Stability Analysis of the Oscillation Phases of a Two-Stage Ring-Oscillator

Equation (4.24) states that when the two-stage ring-oscillator of Fig. 4.3 is free-running ($I_{inj} = 0$), the outputs have a phase difference of $\Delta\theta = \pm\pi/2$. In this section we use perturbation analysis, a similar approach to [15], to investigate the stability of these solutions for $\Delta\theta$. We first start with $\Delta\theta = +\pi/2$. For this case $\theta_{1,1}$ and $\theta_{2,1}$ can be expressed as

$$\theta_{1,1}(t) = \omega_{\mathrm{SRF}}t + \delta\theta_1 \tag{A.1a}$$

$$\theta_{2,1}(t) = \omega_{\mathrm{SRF}}t - \frac{\pi}{2} + \delta\theta_2 \tag{A.1b}$$

where $\delta\theta_1$ and $\delta\theta_2$ are perturbations added to $\theta_{1,1}$ and $\theta_{2,1}$ respectively. Substituting (A.1a) and (A.1b) into (4.21a) and (4.21b) results in

$$\omega_{\mathrm{SRF}} + \frac{\mathrm{d}}{\mathrm{d}t}\delta\theta_1(t) = \frac{1}{RC}\frac{I_1\cos(\delta\theta_1(t) - \delta\theta_2(t))}{I_2 + I_1\sin(\delta\theta_1(t) - \delta\theta_2(t))} \tag{A.2a}$$

$$\omega_{\mathrm{SRF}} + \frac{\mathrm{d}}{\mathrm{d}t}\delta\theta_2(t) = \frac{1}{RC}\frac{I_1\cos(\delta\theta_1(t) - \delta\theta_2(t))}{I_2 - I_1\sin(\delta\theta_1(t) - \delta\theta_2(t))} \tag{A.2b}$$

If we define $\Delta(\delta\theta)$ as $\delta\theta_1 - \delta\theta_2$, and also considering that $\delta\theta_1$ and $\delta\theta_2$ are very small compared to $\theta_{1,1}$ and $\theta_{2,1}$, we can derive a differential equation for $\Delta(\delta\theta)$ using (A.2a) and (A.2b), as shown below.

$$\frac{\mathrm{d}}{\mathrm{d}t}\Delta(\delta\theta(t)) \approx -\frac{1}{\tau}\Delta(\delta\theta(t)) \tag{A.3}$$

where

$$\tau = \frac{1}{2\omega_{\mathrm{SRF}}}\frac{I_2}{I_1}. \tag{A.4}$$

The solution to (A.3) is

M. Farazian et al., *Fast Hopping Frequency Generation in Digital CMOS*,
DOI: 10.1007/978-1-4614-0490-3,

$$\Delta(\delta\theta(t)) = \Delta(\delta\theta(0))\exp(-t/\tau). \tag{A.5}$$

As can be seen from (A.5), any perturbation on the phase difference will eventually diminish. A similar analysis can be done for $\delta\theta_1(t)$ and $\delta\theta_2(t)$. From (A.2a), (A.2b), and (A.5) the solution for $\delta\theta_1(t)$ and $\delta\theta_2(t)$ can be expressed as follows:

$$\delta\theta_1(t) = \frac{I_2}{I_1}\frac{1}{2\tau}\left[\tau\ln\left(e^{t/\tau} + \frac{I_1}{I_2}\Delta(\delta\theta(0))\right) - t\right] \tag{A.6a}$$

$$\delta\theta_2(t) = \frac{I_2}{I_1}\frac{1}{2\tau}\left[\tau\ln\left(e^{t/\tau} - \frac{I_1}{I_2}\Delta(\delta\theta(0))\right) - t\right] \tag{A.6b}$$

Using (A.6a) and (A.6b) it can be shown that $\lim_{t\to\infty}\delta\theta_1(t) = 0$ and $\lim_{t\to\infty}\delta\theta_2(t) = 0$.

A similar analysis for $\Delta\theta = -\pi/2$ results in

$$\Delta(\delta\theta(t)) = \Delta(\delta\theta(0))\exp(+t/\tau) \tag{A.7}$$

which shows that, in this case, any perturbation sustains and grows with time. A similar analysis can be performed to check the stability of the solutions for oscillation phases in the presence of an external signal (I_{inj}).

Appendix B
Step Response of Injection-Locked Two-Stage Ring-Oscillator

Assuming that the ring-oscillator of Fig. 4.8 is injection locked to a signal at frequency ω_{inj} and steady-state is reached. We also assume that the conditions stated in Sect. 4.5.3 for quadrature equal amplitude output voltages ($V_{a1} = V_{a2}$ and $\Delta\theta = \pi/2$) are satisfied, i.e., $I_{\text{inj1}} = I_{\text{inj2}} = I_{\text{inj}}$, $\psi_1 = \psi_2 = \psi$. Hence, (4.21a) results in

$$\omega_{\text{inj}} = \frac{1}{RC}\frac{I_1 + \frac{\pi}{4} I_{\text{inj}} \sin\psi}{I_2 + \frac{\pi}{4} I_{\text{inj}} \cos\psi}. \tag{B.1}$$

If a phase step with magnitude of $\Delta\theta_{\text{inj}}$ is applied to both inputs to the oscillator at time $t = 0^+$, i.e., both θ_{inj1} and θ_{inj2} jump for $\Delta\theta_{\text{inj}}$ from their initial values, $\theta_{1,1}(t)$ and $\theta_{2,1}(t)$ for $t > 0^+$ would change accordingly as

$$\theta_{1,1}(t) = \omega_{\text{inj}} t + \Delta\theta_{1,1}(t) \tag{B.2a}$$

$$\theta_{2,1}(t) = \omega_{\text{inj}} t - \pi/2 + \Delta\theta_{2,1}(t). \tag{B.2b}$$

Consequently, $\psi_1(t)$ and $\psi_2(t)$ for $t > 0^+$ are

$$\psi_1(t) = \psi + \Delta\theta_{\text{inj}} - \Delta\theta_{1,1}(t) \tag{B.3a}$$

$$\psi_2(t) = \psi + \Delta\theta_{\text{inj}} - \Delta\theta_{2,1}(t). \tag{B.3b}$$

In addition, $\Delta\theta(t)$ undergoes the following change.

$$\Delta\theta(t) = \pi/2 + \Delta\theta_{1,1}(t) - \Delta\theta_{2,1}(t) \tag{B.4}$$

M. Farazian et al., *Fast Hopping Frequency Generation in Digital CMOS*,
DOI: 10.1007/978-1-4614-0490-3,

Assuming $\Delta\theta_{1,1}(t) - \Delta\theta_{2,1}(t)$ is negligible compared to $\pi/2$, and substituting (B.2a) and (B.3a) into (4.21a) results in

$$\omega_{\text{inj}} + \frac{\mathrm{d}}{\mathrm{d}t}\Delta\theta_{1,1}(t) = \frac{1}{RC}\frac{I_1 + \frac{\pi}{4}I_{\text{inj}}\sin\psi_1(t)}{I_2 + \frac{\pi}{4}I_{\text{inj}}\cos\psi_1(t)}. \tag{B.5}$$

By substituting (B.3a) into (B.5) and assuming that $\Delta\theta_{\text{inj}} - \Delta\theta_{1,1}(t)$ is small compared to ψ we obtain

$$\begin{aligned}\omega_{\text{inj}} + \frac{\mathrm{d}}{\mathrm{d}t}\Delta\theta_{1,1}(t) = &\frac{1}{RC}\frac{I_1 + \frac{\pi}{4}I_{\text{inj}}\sin\psi}{I_2 + \frac{\pi}{4}I_{\text{inj}}\left[\cos\psi - \left(\Delta\theta_{\text{inj}} - \Delta\theta_{1,1}(t)\right)\sin\psi\right]}\\ &+\frac{1}{RC}\frac{\frac{\pi}{4}I_{\text{inj}}\left(\Delta\theta_{\text{inj}} - \Delta\theta_{1,1}(t)\right)\cos\psi}{I_2 + \frac{\pi}{4}I_{\text{inj}}\left[\cos\psi - \left(\Delta\theta_{\text{inj}} - \Delta\theta_{1,1}(t)\right)\sin\psi\right]}.\end{aligned} \tag{B.6}$$

Since $\frac{\pi}{4}I_{\text{inj}}\left(\Delta\theta_{\text{inj}} - \Delta\theta_{1,1}(t)\right)$ is smaller than I_2, we can simplify the denominator in (B.5). Using (B.1), Eq. (B.6) results in

$$\frac{\mathrm{d}}{\mathrm{d}t}\Delta\theta_{1,1}(t) = -\frac{1}{\tau}\left(\Delta\theta_{1,1}(t) - \Delta\theta_{\text{inj}}\right) \tag{B.7}$$

where

$$\tau = \frac{I_2 + \frac{\pi}{4}I_{\text{inj}}\cos\psi}{\frac{\pi}{4}I_{\text{inj}}\cos\psi}RC. \tag{B.8}$$

Using (B.1), Eq. (B.8) can be rewritten as

$$\tau = \frac{I_1 + \frac{\pi}{4}I_{\text{inj}}\sin\psi}{\frac{\pi}{4}I_{\text{inj}}\omega_{\text{inj}}\cos\psi}. \tag{B.9}$$

Equation (B.7) shows that an injection-locked two-stage ring-oscillator tracks the step on the phase of the injection signal with a time constant τ. Consequently, $\Delta\theta_{1,1}(t)$ approaches $\Delta\theta_{\text{inj}}$. A similar conclusion is obtained for $\Delta\theta_{2,1}(t)$ following the same steps, i.e.,

$$\frac{\mathrm{d}}{\mathrm{d}t}\Delta\theta_{2,1}(t) = -\frac{1}{\tau}\left(\Delta\theta_{2,1}(t) - \Delta\theta_{\text{inj}}\right). \tag{B.10}$$

As a result, the final values of $\theta_{1,1}(t)$ and $\theta_{2,1}(t)$ can be expressed as

$$\theta_{1,1}(t) \to \omega_{\text{inj}}t + \Delta\theta_{\text{inj}} \tag{B.11a}$$

$$\theta_{2,1}(t) \to \omega_{\text{inj}}t + \Delta\theta_{\text{inj}} - \frac{\pi}{2}. \tag{B.11b}$$

Equations (B.7) and (B.10) represent first-order systems with a transfer function

$$G(S) = \frac{1}{1 + S/\omega_P} \tag{B.12}$$

where

$$\omega_P = \frac{1}{\tau} = \frac{\frac{\pi}{4} I_{\text{inj}} \cos \psi}{I_1 + \frac{\pi}{4} I_{\text{inj}} \sin \psi} \omega_{\text{inj}}. \tag{B.13}$$

Appendix C
Analysis of the Negative Impedance Generator

The negative impedance generator used in the amplifier of Fig. 5.14b is redrawn in Fig. C.1a. We start with the small-signal analysis of this circuit. The input admittance of the circuit shown in Fig. C.1a is given by (5.17). $Y_{\mathrm{N}}(s)$ consists of a parallel capacitor, and a residual part, $Y_0(s)$, which is given by (C.1).

$$Y_0(s) = -sC_{\mathrm{N}} \frac{g_{m2} - sC_{\mathrm{gs2}}}{g_{m2} + s(C_{\mathrm{gs2}} + 2C_{\mathrm{N}})} \tag{C.1}$$

Equation (C.1) shows that Y_0 has a net capacitive part, which can be found as

$$\lim_{s\to\infty} \frac{Y_0}{s} = \frac{C_{\mathrm{gs2}}C_{\mathrm{N}}}{C_{\mathrm{gs2}} + 2C_{\mathrm{N}}} \tag{C.2}$$

The term given by (C.2) is a series combination of two capacitors with the values of $C_{\mathrm{gs2}}/2$ and C_{N}. Therefore Y_0 can be written as

$$Y_0(s) = \frac{C_{\mathrm{gs2}}C_{\mathrm{N}}}{C_{\mathrm{gs2}} + 2C_{\mathrm{N}}} s + Y_1(s) \tag{C.3}$$

It can be shown that Y_1 is a series combination of a resistor and a capacitor. So, Z_1 can be written as

$$Z_1(s) = \frac{1}{Y_1(s)} = R_1 + \frac{1}{sC_1} \tag{C.4}$$

where R_1 and C_1 are given by

$$R_1 = -\frac{(2C_{\mathrm{N}} + C_{\mathrm{gs2}})^2}{2C_{\mathrm{N}}(2C_{\mathrm{N}} + C_{\mathrm{gs2}})g_{m2}} \tag{C.5a}$$

$$C_1 = -\frac{(2C_{\mathrm{N}} + C_{\mathrm{gs2}})^2}{2C_{\mathrm{N}}(2C_N + C_{\mathrm{gs2}})}. \tag{C.5b}$$

M. Farazian et al., *Fast Hopping Frequency Generation in Digital CMOS*,
DOI: 10.1007/978-1-4614-0490-3,

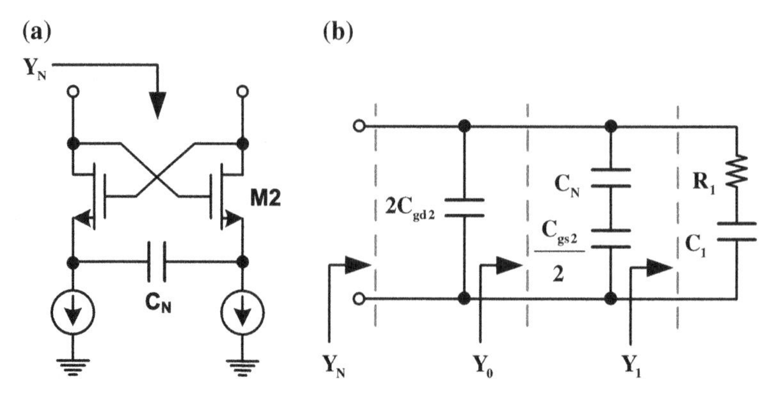

Fig. C.1 **a** Negative impedance generator, **b** equivalent circuit for negative impedance generator

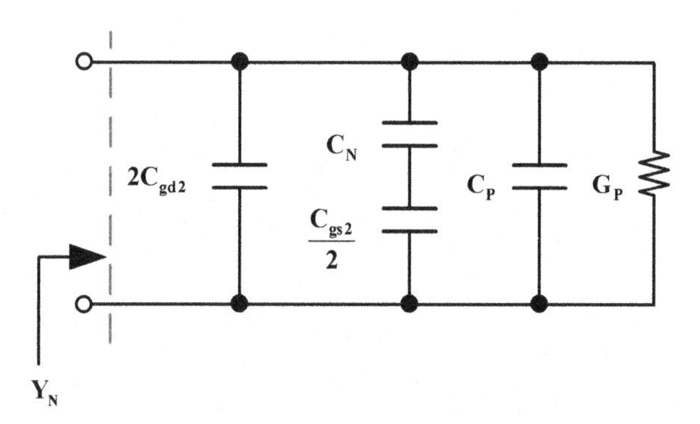

Fig. C.2 Equivalent circuit for negative impedance generator

Figure C.1b shows an equivalent circuit for the negative impedance generator of Fig. C.1a in which R_1 and C_1 are independent of frequency. However, a parallel combination of R_1 and C_1, as shown in Fig. C.2, is more desirable for analysis of the net capacitance and conductance of this circuit.

To do so, we define $Y_1(s)$ as

$$Y_1(s) = G_P + sC_P \tag{C.6}$$

where G_P and C_P are given by

$$G_P = -\frac{2C_N(C_N + C_{gs2})\omega^2 g_{m2}}{g_{m2}^2 + (2C_N + C_{gs2})^2\omega^2} \tag{C.7a}$$

$$C_P = -\frac{2C_N(C_N + C_{gs2})}{2C_N + C_{gs2}} \cdot \frac{g_{m2}^2}{g_{m2}^2 + (2C_N + C_{gs2})^2\omega^2}. \tag{C.7b}$$

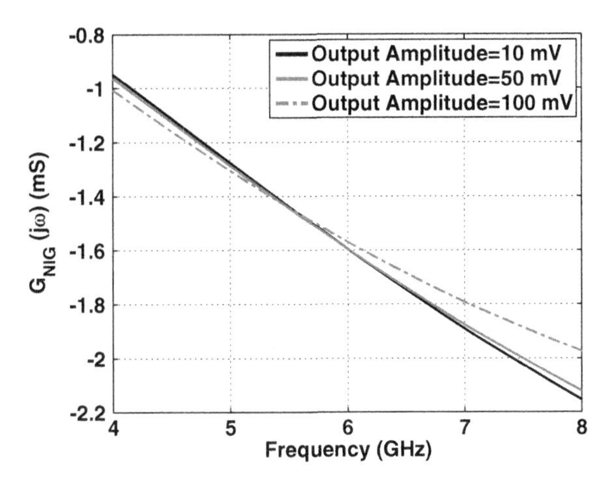

Fig. C.3 Simulated G_{NIG} versus frequency for different values of output amplitude

Equations (C.7a) and (C.7b) show that the circuit of Fig. C.1a can exhibit negative capacitance and negative conductance, which lead to gain and bandwidth expansion at high frequencies.

The negative capacitance obtained from the circuit of Fig. C.1a becomes negligible in the large-signal regime. However, the negative conductance has a small variation over a wide range of signal swings. In this case, if we assume that the total output conductance and capacitance of the differential pair of Fig. 5.14b, without the negative impedance generator, are G_{L} and C_{L} respectively, and the total output conductance and capacitance of the negative impedance generator of Fig. C.1a are respectively $G_{\mathrm{NIG}}(S)$ and $C_{\mathrm{NIG}}(S)$, then the transfer function of the amplifier of Fig. 5.14b can be written as

$$A(S) = \frac{G_m}{G_{\mathrm{eq}}(S)} \cdot \frac{1}{1 + SC_{\mathrm{eq}}(S)/G_{\mathrm{eq}}(S)} \tag{C.8}$$

where G_m is the large signal transconductance of transistor $M1$ in Fig. 5.14b and

$$G_{\mathrm{eq}}(S) = G_{\mathrm{L}} + G_{\mathrm{NIG}}(S) \tag{C.9a}$$

$$C_{\mathrm{eq}}(S) = C_{\mathrm{L}} + C_{\mathrm{NIG}}(S). \tag{C.9b}$$

Figure C.3 shows the simulated G_{NIG} versus frequency for various output swings. As can be seen from Fig. C.3, G_{NIG} within the frequencies of interest (6–8 GHz) can be approximated by

$$G_{\mathrm{NIG}}(j\omega) = -g_0 - g_1\omega \tag{C.10}$$

where g_0 and g_1 are positive numbers. From (C.8) and (C.10) we obtain

$$|A(j\omega)| = \frac{G_m}{\left[\left(C_{\text{eq}}^2(j\omega) + g_1^2\right)\omega^2 - 2(G_{\text{L}} - g_0)g_1\omega + (G_{\text{L}} - g_0)^2\right]^{\frac{1}{2}}}. \tag{C.11}$$

The maximum value of the transfer function described in (C.11) occurs at $\omega = \widetilde{\omega}$ where $\widetilde{\omega}$ is given by

$$\widetilde{\omega} = (G_{\text{L}} - g_0)g_1/(\overline{C}_{\text{eq}}^2 + g_1^2) \tag{C.12}$$

and $\overline{C}_{\text{eq}}$ is the average value of $C_{\text{eq}}(j\omega)$ at the vicinity of $\widetilde{\omega}$. By evaluating (C.11) at $\omega = \widetilde{\omega}$ we obtain

$$|A(j\widetilde{\omega})| = \frac{G_m}{|G_{\text{L}} - g_0|}\left[1 + \left(g_1/\overline{C}_{\text{eq}}\right)^2\right]^{\frac{1}{2}}. \tag{C.13}$$

The gain given by (C.13) can be significantly larger than the DC gain of the differential pair without the negative impedance generator ($G_{\text{m}}/G_{\text{L}}$).

Appendix D
Zero Frequencies of a Multi-Stage Polyphase Filter

Figure D.1a shows a multi-stage polyphase filter that is implemented by cascading N single-stage passive polyphase filters. Each stage can be implemented using the single-stage polyphase filter of Fig. 6.6a or b. As can be seen from (6.7) and (6.10), both polyphase filters of Fig. 6.6a and b achieve an imaginary zero at

$$s = \frac{\pm j}{R_{z1}C_{z1}} \tag{D.1}$$

when quadrature input phases are applied. Also, as discussed earlier, the LHP or the RHP zero depends on whether clockwise or counterclockwise input phases are used. Consequently, the kth stage in the cascade of Fig. D.1a has a zero at $\pm j/R_{zk}C_{zk}$, where R_{zk} and C_{zk} are the values of the resistors and the capacitors that are used to implement the kth stage.

In order to find the zero frequencies of the cascade of Fig. D.1a we assume that the last stage in this cascade is implemented using the single-stage passive filter of Fig. D.1b. The direct way of finding the zero frequencies of the N-stage cascade of polyphase filter of Fig. D.1a is by deriving its transfer function. However, here we use a different approach. At the zero frequencies, the outputs of the polyphase filter of Fig. D.1a are zero regardless of its quadrature inputs, i.e.,

$$V_{O_{I+}} = V_{O_{Q+}} = V_{O_{I-}} = V_{O_{Q-}} = 0. \tag{D.2}$$

In this case, applying (6.5) to the Nth stage in the cascade of Fig. D.1a, results in the following relationships between the inputs to the Nth stage.

$$V_{\mathrm{A}} + sR_{z\mathrm{N}}C_{z\mathrm{N}}V_{\mathrm{B}} = 0 \tag{D.3a}$$

$$V_{\mathrm{B}} + sR_{z\mathrm{N}}C_{z\mathrm{N}}V_{\mathrm{C}} = 0 \tag{D.3b}$$

$$V_{\mathrm{C}} + sR_{z\mathrm{N}}C_{z\mathrm{N}}V_{\mathrm{D}} = 0 \tag{D.3c}$$

M. Farazian et al., *Fast Hopping Frequency Generation in Digital CMOS*,
DOI: 10.1007/978-1-4614-0490-3, © Springer Science+Business Media New York 2013

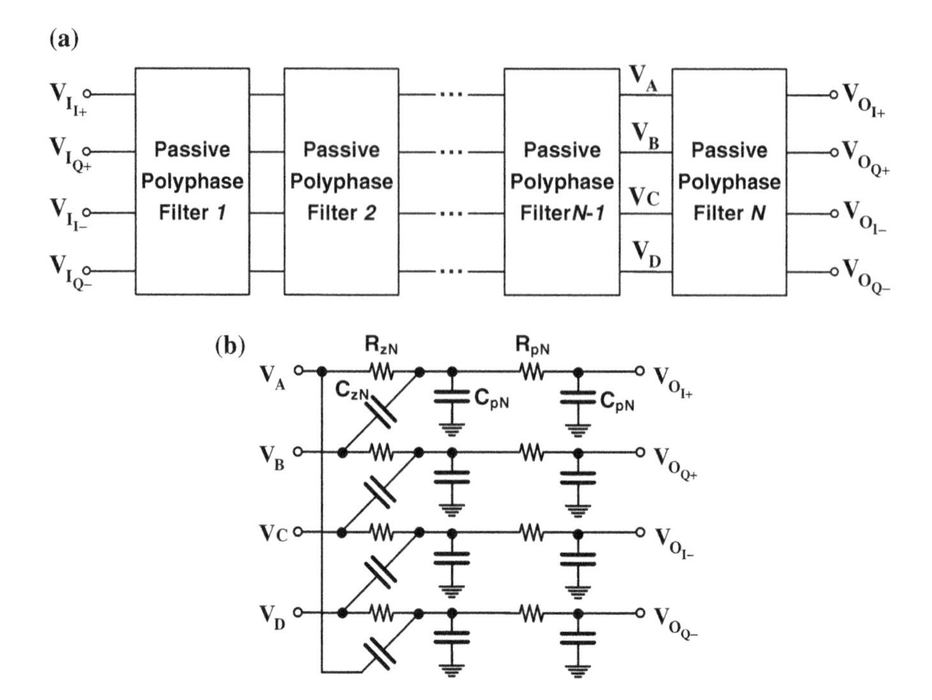

Fig. D.1 **a** Cascade of *N* single-stage polyphase filter, **b** schematic of the *N*th stage

$$V_{\mathrm{D}} + sR_{zN}C_{zN}V_{\mathrm{A}} = 0 \tag{D.3d}$$

Solving (D.3) for V_{A} leads to

$$V_{\mathrm{A}}\left(1 - (sR_{zN}C_{zN})^4\right) = 0. \tag{D.4}$$

Solution 1

If V_{A} in (D.4) is nonzero, the frequency of a zero corresponding to the *N*th stage is at $\pm 1/R_{zN}C_{zN}$ or $\pm j/R_{zN}C_{zN}$. The first solution ($\pm 1/R_{zN}C_{zN}$) is not an imaginary zero, and also does not correspond to quadrature input phases to the *N*th stage in Fig. D.1a. However, the second solution ($\pm j/R_{zN}C_{zN}$) is an imaginary zero and is obtained only when the inputs to the *N*th stage have a clockwise or a counterclockwise quadrature sequence. Consequently, the imaginary zero of the *N*th stage happens at

$$s = \frac{\pm j}{R_{zN}C_{zN}}. \tag{D.5}$$

From (D.5) it can be observed that cascading *N* polyphase filters does not change the frequency of the imaginary zero that corresponds to the last stage.

Solution 2

The second solution to (D.4) is $V_A = 0$. In this case, from (D.3) it is concluded that

$$V_A = V_B = V_C = V_D = 0. \quad (D.6)$$

The solution to (D.6) gives the zero frequencies of the $(N - 1)$th stage in the cascade of Fig. D.1a. Using a similar approach to (D.3)–(D.5) it can be concluded that (D.6) corresponds to an imaginary zero at $\pm 1/R_{z(N-1)}C_{z(N-1)}$, or to a zero input to the $(N - 1)$ th stage in Fig. D.1a.

This technique can be used recursively to find all the imaginary zeros of the N-stage polyphase filter of Fig. D.1a. It can be concluded that the N-stage polyphase filter of Fig. D.1a has N imaginary zeros that are given by

$$Z_k = \pm j/R_{zk}C_{zk} \quad k = 1, 2, \ldots, N. \quad (D.7)$$

As can be seen from (D.7), when cascading the polyphase filters of Fig. 6.6a and b—even without inserting a buffer between the stages—the location of the imaginary zeros is preserved.

Index

M. Farazian et al., *Fast Hopping Frequency Generation in Digital CMOS*,
DOI: 10.1007/978-1-4614-0490-3,